Geometry in Action

A Discovery Approach Using The Geometer's Sketchpad®

Clark Kimberling

University of Evansville

Key College Publishing
Innovators in Higher Education

www.keycollege.com

Clark Kimberling
University of Evansville
Department of Mathematics
1800 Lincoln Avenue
Evansville, IN 47722
ck6@evansville.edu

Key College Publishing was founded in 1999 as a division of Key Curriculum Press® in cooperation with Springer-Verlag New York, Inc. We publish innovative texts and courseware for the undergraduate curriculum in mathematics and statistics as well as mathematics and statistics education. For more information, visit us at www.keycollege.com.

Key College Publishing
1150 65th Street
Emeryville, CA 94608
(510) 595-7000
info@keycollege.com
www.keycollege.com

Geometry in Action: Selected Sketches CD-ROM
Key College Publishing guarantees that the CD-ROM that accompanies this book is free of defects in materials and workmanship. A defective disk will be replaced free of charge if returned within 90 days of the purchase date. After 90 days, there is a $10.00 replacement fee.

Development Editors: Jacqueline Meijer-Irons, Allyndreth Cassidy
College Production Project Manager: Michele Julien
Copyeditor: Balwant Khalsa
Production Director: Diana Jean Parks
Text Designer: Adriane Bosworth
Compositor: ICC
Art and Design Coordinator: Kavitha Becker
Cover Designer: Todd Michael Bushman
Prepress: Versa Press
Printer: Versa Press

Executive Editor: Richard Bonacci
General Manager: Mike Simpson
Publisher: Steven Rasmussen

Library of Congress Cataloging-in-Publication Data

Kimberling, Clark, 1942–
   Geometry in action: a discovery approach using The Geometer's Sketchpad / Clark Kimberling.
     p. cm.
   Includes bibliographical references and index.
   ISBN 1-931914-02-8 (pbk.)
    1. Geometry. 2. Geometer's Sketchpad. I. Title.

QA445 .K556 2003
516–dc21
                                    2002030186

Printed in the United States of America
10 9 8 7 6 5 4 3 2   06 05

Brief Contents

In addition to the Detailed Contents on the following pages, you will find a very useful appendix, Assignments by Filename, at the end of the book, on pages 177–185.

Take a glance ahead to page 12, and note that each assignment has a name, such as **SS1A** and **SS1B**. These names will also serve as filenames for assigned sketches. The appendix gives a complete list of filenames with brief descriptions.

Detailed Contents

Preface

Geometry in Action: A Discovery Approach Using The Geometer's Sketchpad®
stems from a remarkably successful geometry classroom experience I had
using The Geometer's Sketchpad software along with a geometry textbook.
The students spent part of each class period *doing* geometry on Sketchpad™.
Seated two per computer, they carried out assignments in the form of original sketches. Through the assignments, Sketchpad enabled the students to
interact directly with geometry itself and also to interact with one another.
As one student said, Sketchpad makes geometry come alive, "like my cat,
I mean, you get instant feedback when you do something right ... and vice
versa".

Why does Sketchpad work so well as an educational tool? The main reason must surely be that the basic Sketchpad commands match the basics of
geometry. Commands like **Construct | Parallel Line** are realizations of Euclid's
postulates. Consequently, developing a sketch is akin to discovering a theorem. Sketchpad puts students in touch with elemental geometry itself,
greatly reducing the language barrier that often comes between students
and geometry.

A second reason that students love to use Sketchpad is that it is a "pad"
for their own creativity. It is often said that creative, discovery-based learning
is the deepest kind. As Sherry Turkle explains in *The Second Self: Computers
and the Human Spirit* (second edition forthcoming from MIT Press), those
who create on a computer *discover themselves* in new ways.

Another reason for students to create their own sketches is the development of their ability to self-express. The precise and deductive character of
geometry makes it an ideal substance for molding ideas and relationships
into words. As a part of their *Geometry in Action* assignments, students

write captions that tell what geometric properties are being illustrated, with instructions for "running" the sketches.

Geometry in Action is intended primarily for three kinds of use:

- as a supplement for geometry courses, especially those taken by prospective teachers
- as a basis for independent study and undergraduate research
- as a modest encyclopedia of standard topics and gems that are well suited to Sketchpad

The topics represented in *Geometry in Action* are chosen according to what works best on Sketchpad. Accordingly, not all the standard topics of a college geometry course are included. For example, you won't find comparisons of postulate collections, and you won't find finite geometries. These topics do not lend themselves to sketching as directly as the selected topics do, although they *are* important topics in college geometry.

The criterion stated above for the topics chosen for *Geometry in Action*—namely, what works best on Sketchpad—has led to the inclusion of two kinds of topics: standard and enrichment. You can easily distinguish between them by comparing the table of contents of *Geometry in Action* with the contents of your textbook. One thing will quickly catch the eye: the first four *Geometry in Action* chapters have Sketchpad names. While these names serve the purpose of advancing the student's facility with Sketchpad, they also lend themselves naturally to a succession of standard geometric topics:

Chapter 1: Simple Sketches introduces points, lines, circles, triangles, classical examples of these, and a first look at their representations using coordinates.

Chapter 2: Tools lends itself to special points of a triangle (centroid, circumcenter, etc.) and the fundamental concepts of perspective, harmonic conjugates, and Ceva's theorem.

Chapter 3: Locus offers a magnificent approach to parabolas, ellipses, and hyperbolas, as well as inversion in a circle and the graphing of functions.

Chapter 4: Animation enables action-packed realizations of translation, reflection, dilation, and similarity. Section 1 typifies another aspect of animation: *continuously updated measurements,* in this case confirming in three ways the area-of-triangle formula, for hundreds of different triangles.

There are several reasons for including and recommending **Chapter 5: Trigonometry.** One, already implied, is that Sketchpad "fits" trigonometry *very* well. Another is that precisely *because* students have already studied trigonometry elsewhere, they *should* study it again. A third reason for including trigonometry is that the closely related hyperbolic functions are needed later in **Chapter 9: Hyperbolic Geometry.** In particular, the functions *cosh* and *sinh* are necessary for understanding the hyperbolic Pythagorean theorem and the hyperbolic laws of sines and cosines. Many students will need to brush up on the "old" laws of sines and cosines in preparation for their hyperbolic cousins. Such a brush-up is provided, for example, by assignments

TR3A and **TR4B** (the naming of assignments and files is explained thoroughly in the appendix).

Geometry in Action is discovery-oriented. That is to say, the material is written to enable students to discover geometric properties on their own, in two ways: writing their own sketches, and using their sketches for assignments, projects, and original explorations. **Chapter 6: Making Your Own Discoveries** aims to cultivate the ability to make original discoveries. Sketchpad is great for this! The chapter proceeds by themes and variations and includes fresh ideas for undergraduate research projects, including new loci associated with rotations, and properties of the mysterious Philo line. There is also a section that explains how good discoveries get published.

Chapter 7: Famous Discoveries contains some of the great nuggets of geometry, including two-column proofs of the Pythagorean theorem and implications of the famous theorem that Pascal discovered as a teenager. **Chapter 8: Selected Topics** includes polar coordinates (which Sketchpad performs admirably), advanced conics (for example, how to construct the center of a conic from five points on the conic), geometric realizations of the golden mean, and animated tessellations.

One idealized discovery-oriented approach would consist of a hand-out on the first day of class, listing postulates and stating a single semester-long assignment: *to discover what the postulates imply*. That approach (known as the Moore method) represents one end of a spectrum of pedagogical possibilities. The other end is the here's-everything-you-should-do approach, which, of course, is not really discovery-learning at all. *Geometry in Action* aims between those extremes with a purposefully concise style. The conciseness allows an openness for the student—the kind of openness that is essential for exploration and discovery-making.

For example, the wording of certain assignments is like this: "Predict what will happen when you drag the points and use your sketch to confirm or refute your predictions". This wording nets a variety of good responses, especially in group-learning situations. What happens is that the students find what they are supposed to find, and then go on to find more. For example, if the points to be dragged are A, B, C, and the orthocenter H of triangle ABC, then some students will also discover that A is the orthocenter of triangle HBC. Some will go on to suspect and then discover that B is the orthocenter of triangle AHC, and so on. To summarize: The payoff of a concise and open style of presentation is that students will receive the content in the form of seeds, ready for germination, growth, and flowering.

An Invitation to the Student

There is so much more to *Geometry in Action* than just what you can read in the book. Now I'd like to invite you to try out the CD that can be found inside the back cover. Upon loading it into your computer, you'll find 27 sketches that show what's really meant by "geometry in *action*".

The first file on the CD is named "An Introduction Read It". Open it, and read the names of the sketches in alphabetical order, along with brief descriptions. Then download **Dancer**. You'll see three stick figures, or humanoids. The middle one is the dancer. Use your mouse to drag point X, and already you'll see action—but there's much more to come! Drag the point labeled "hand". Then drag point h to see how this point determines where "hand" can move. Now for the big step: Select point X (by clicking on it), then click **Display** at the top of the screen, and then click **Animate** in the **Display** menu. Now the dancer slides back and forth automatically. During the motion, you can drag the points that determine the shape of the dancer. In fact, you can select the "hand" and "knee" and animate them, too, while the main animation continues!

This initial sketch is just a taste of what is to come and hopefully it has piqued your interest. You are now ready to explore further the wonderful world of The Geometer's Sketchpad.

Acknowledgements

I would like to thank those who read the manuscript in various stages. Thank you to the reviewers: Roger Engle, Clarion University; Helen Gerretson, University of Northern Colorado; John Goulet, Worcester Polytechnic University; Charlie Jacobson, Elmira College; John Olive, University of Georgia; David Royster, University of North Carolina, Charlotte; Alan Russell, Elon University; and Paul Yiu, Florida Atlantic University.

I am very grateful to the people at Key College Publishing and KCP Technologies for advice and patience during the preparation of *Geometry in Action*. Thank you, Richard, Jacqueline, Allyndreth, Keith, Scott, Steven, and Nick.

Finally, I want to thank the geometry class that tried out a preliminary version of *Geometry in Action*. Thanks, Kimberly, Jason, Michael, Amy, Stacy, Joanne, Johonna, Terrynn, Marla, Katrina, . . . You will recall the evenings when we talked about the outstanding work you were doing and the possibility that the sketches and notes would someday grow into a book. Thanks for your ideas and enthusiasm, which are spread across the pages that follow.

Clark Kimberling
University of Evansville

Sketchpad Basics

THE OBJECTIVE IN these preliminary pages is to explore The Geometer's Sketchpad® in such a way as to quickly learn some basics. Within an hour, three things are likely to happen:

- You'll create some amazing results.
- You'll develop a desire to go beyond the basics.
- You'll click **Help**.

The Toolbox

When Sketchpad™ opens, six icons appear on the upper left side of the screen, along with words across the top. Click **File** to view a dropdown menu. On it, click **New Sketch**. The way to write such a combination is this: **File** | **New Sketch**.

The second icon on the toolbox of six icons has a dot in the middle. Click it. On the screen there's an arrow that will follow your guidance of the mouse. Stop the arrow and left-click the mouse. You should see a point on the screen. If it doesn't have a label, apply **Edit** | **Preferences** | **Text**, and follow instructions so that new points will be automatically labeled.

Click several more points onto the screen. Notice that each time you add a point, it stays highlighted until you add another. This highlighting indicates that the point is *selected*—a very important term.

Frequently, you'll need to select and de-select objects. To select an object, first click the **Selection Arrow** tool, at the top of the toolbox. Then place the mouse-guided arrow so that its tip is right on the object to be selected, and click. Clicking again de-selects the object.

After selecting an object, it will (usually) stay selected until you de-select it. Aside from points, other objects are lines, circles, and segments. To see a line, for example, select two points and apply **Construct | Line**.

Next, we want to construct a circle, but first let's note a common oversight: leaving objects selected when you shouldn't. Without de-selecting your line (which was automatically selected when it was constructed), select two points, and try to apply **Construct | Circle By Center+Point**. It won't work because your selected objects are not the right "givens". So, de-select the line, select the two points, and construct a circle.

Select your line and apply **Display | Line Width**. Try all three options. These apply to segments, rays, circles, and loci—which we'll sketch later.

Now click on the fifth tool in the toolbox, called **Text** and labeled **A**. The mouse-controlled arrow changes to a hand with a pointing finger. Move it so that the finger nearly touches the bottom of a point. The hand will darken. Left-click your mouse to hide the label. Click again to restore the label.

Next, move the hand so that it darkens again. Double-click rapidly to view a dialogue box for changing the label. Practice changing the character, font, style, and size. Practice unhiding and hiding the label of your line and circle.

If any of your attempts with labels failed, it may be that you clicked when the hand was not close enough to the object. Do that intentionally now—a quick double-click. The result is a small rectangle. As you type characters into it, the rectangle will grow to accommodate them. Type your name; and then click the **Arrow** tool. The arrow replaces the hand, and with the arrow you can select the **caption** you typed and drag it to a different location. When finished, be sure to de-select it. Any time you want to edit or add to a caption, click on the **Text** tool, and then click within the caption where you wish to start. While editing, you'll see at the bottom of the screen some darkened buttons (If not, apply **Display | Show Text Palette**.) These enable you to apply boldface, italics, underlining, font changes, and symbols. Try out those buttons, and verify that you can put several captions on the screen.

Deleting objects, including captions, is easy. Select the object and tap the **Delete** key. Sometimes, objects get deleted by mistake. In that case—or any other time that you want to "go back"—press the Control key (**Ctrl** on Windows, ⌘ on Macintosh) and, keeping it down, tap the **Z** key. Repeating this takes you back several steps. Another way to go back is to apply **Edit | Undo**.

Next, put six more points on the screen, and apply **Construct | Segment** to two of them. Then, imagine a rectangle just large enough to contain your segment. Place the arrow on the upper left corner of the imaginary rectangle, press and hold the mouse button, moving the arrow diagonally across your imaginary rectangle. Release, noting that the objects inside the rectangle are now selected as a "block". The rectangle vanishes, and now there are several things you can do:

- drag the block
- apply **Edit | Copy**, followed by **Edit | Paste**, followed by dragging the pasted block away from the original
- hide the block by applying **Display | Hide Objects**
- delete the block using the **Delete** key

Practice all four, and practice using **Ctrl Z** (Windows) or ⌘ **Z** (Mac) to "go back".

So far, we've used three of the six tools, namely **Selection Arrow**, **Point**, and **Text**. The third tool in the toolbox is called **Compass**, and the fourth, **Straightedge**. It is easy to understand how these work, and much of what these tools do can be done just as well using the **Construct** menu instead. The final tool, called **Custom Tools**, is like a key to a potentially enormous repository of further tools. We'll consider **Custom Tools** in Section 7.

SECTION 2   Construct

You've already used **Construct** to create segments, lines, and circles. In order to sample the **Construct** menu further, start with the following objects on your screen: several points, a segment, a line, and a circle. Use the **Text** tool to label the endpoints of the segment as A and B, two points on the line as C and D, and the center of your circle as O. Note that another point came with your circle; by dragging it you can vary the radius; label this point R.

Click the **Point** tool, place the tip of the arrow directly on line CD, and click. The result is a new point. You can get the same result by applying **Construct | Point On Object**. In any case, label the point P. Click on the **Selection Arrow** and drag P. Notice that P moves *only* on the line.

Put a point on segment AB. Then put a point on your circle, and label it Q. In each case, check that the new point is free to move only on the object you put it on.

Sketchpad enables three kinds of points: *independent, movable,* and *constructed*. Independent points, such as A and B, have no parents—you can drag them with two-dimensional freedom. Movable points, like P and Q, have a parent, namely the object to which they are confined; they have one-dimensional freedom. Constructed points, such as the midpoint of a segment, are completely determined and have zero-dimensional freedom.

In order to further clarify the three kinds of points, select point A and apply **Display | Animate Point**. After observing the result, apply **Display | Stop Animation**. Then give P and Q the same treatment. Can Sketchpad animate points simultaneously? Try it.

Next, select segment *AB*, and then apply **Construct | Midpoint**. (Note that the selection must be the segment, not its endpoints.) Just for practice, sketch two more points, and then sketch their midpoint.

Add two more lines to your screen. Drag one to cross the other, and apply **Construct | Intersection**. Or, you can often simply put the tip of the arrow right where two lines meet, then click, and Sketchpad will sketch the point of intersection. Why not always do it this second way? One answer is that sometimes points may be very close together, making it difficult to select the one you want.

If your circle doesn't meet a line, drag point *R* to create an overlap. Then select the line and the circle, and apply **Construct | Intersections**. Or, if you aren't in a tight situation, you can simply click each point of intersection and get the same results, without using **Construct**. This is especially handy when you want only one of two points of intersection. All of these options apply also to two overlapping circles, two tangent circles, or a line tangent to a circle, although these last two are, unsurprisingly, iffy (so be prepared to find alternatives).

The next few items on the **Construct** menu are easy to use, once you know the necessary "givens":

Parallel Line: Select a line and a point not on the line.

Perpendicular Line: Select a line and a point not on the line.

Angle Bisector: Select three points.

Circle by Center+Point: Select two points.

Circle by Center+Radius: Select a point and a segment.

Arc On Circle: Select a circle and two points on the circle.

Arc Through 3 Points: Select three points.

There are two items remaining on the **Construct** menu. After experimenting with the seven just listed, apply **Edit | Select All** and tap the **Delete** key. Then put three points on the screen. Apply **Edit | Select All** followed by **Construct | Segment**. Drag these points to form a large triangle. De-select all objects (simultaneously) and then select the three points. Apply **Construct | Triangle Interior**. De-select this interior by clicking on it.

Next, mimic triangle-interior construction on a collection of four or more points. One thing to be changed is this: if any segment crosses another segment, you must drag points so that a single polygon is determined.

The final offering under **Construct**, called **Locus**, is a source of great delight. Strictly speaking, the word locus means a path of a point that moves according to some rules. However, Sketchpad enables you to see the "locus" of a segment—or line or circle—as well as the classical kind of locus. Let's start with the locus of a point:

EXAMPLE 1   Place a movable point *P* on a circle, an independent point *Q* outside the circle, and sketch the midpoint *M* of segment *PQ*. Select *P* and *M*, in that order, and apply **Construct | Locus**. You'll see the locus of M. Delete it.

EXAMPLE 2 Continuing, select P and segment PQ, in that order, and apply **Construct | Locus**. This time you'll see the locus of a segment. You can replace segment PQ by line PQ and repeat the steps to see the locus of a line.

EXAMPLE 3 Continuing, sketch the circle centered at P and passing through Q; select P and your new circle, and apply **Construct | Locus**. The result should resemble Figure A.

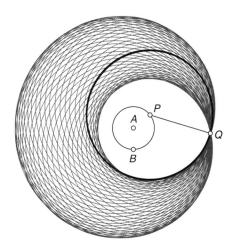

FIGURE A Locus of a circle, as in Example 3

It is highly appropriate, when applying **Locus**, to realize that you are dealing with a function—in the mathematical sense—and that you control the variable and the "rule". Consider the familiar notation $y = f(x)$: in Example 3, the variable x is the point P, the "rule" f is the circle constructed from P, and the domain of f is the circle that P moves on.

SECTION 3 # Display

The **Display** menu offers an option, called **Animate**, that we'll sample in this section. Start with any one of the locus constructions at the end of Section 2. Be sure no object is selected. Then select P and apply **Display | Animate Point**. If a box labeled **Motion Controller** doesn't appear, then apply **Display | Show Motion Controller**. Try out the various controls. You'll see how to stop, start, reverse, and regulate speed. The label "Target: All motions" suggests that Sketchpad can perform several animations simultaneously. To verify this, select Q and apply **Display | Animate Point**. (You needn't stop the animation of P.) Since Q is not confined to a path, as P is, its motion is "random".

Try another quick animation—sketch two circles, put a moving point on each, construct the segment between the two points, and then animate both points. Finally, confirm that you can drag objects on the screen during the motion.

SECTION 4 Measure and Graph

Sketch a circle and place on it three movable points, labeled A, B, C. Sketch their triangle and its interior. Apply **Measure** to each of these: distance from A to B, length of segment AB, circumference of the circle, area of $\triangle ABC$, and area of the circle. You can apply **Edit | Preferences** to change units for future measurements.

In order to sample **Measure | Calculate**, start by applying **Graph | Define Coordinate System**, and observe the effects of dragging the points at 0 and 1 on the horizontal axis. Sketch a circle and place a movable point P on it. Select P and apply **Measure | Abscissa**. Then select P and apply **Measure | Ordinate**. Then apply **Measure | Calculate**. Click on the printed measure of the Abscissa, type + (or click + on the onscreen calculator), then click on the printed measure of the Ordinate, and then click OK. This example illustrates *the purpose of* **Calculate**: *to form functions of variables obtained from sketches.* Very promising!

To continue with the example, select the printed Abscissa and the printed sum that you calculated. Then apply **Graph | Plot As (x,y)**. Label the result R. Finally—for a result that may surprise you—select P and R, in that order, and apply **Construct | Locus**.

SECTION 5 Transform

The **Transform** menu offers four basic transformations: **Translate, Rotate, Dilate, Reflect**. Preceding these on the menu are the means for specifying exactly how you want your transformation to be carried out. For example, to translate a selected object, you must tell Sketchpad the desired direction and distance; or to rotate an object, you must indicate the point about which rotation is to take place and the angle of rotation.

EXAMPLE 4 Sketch an xy coordinate system, and use parallels and perpendiculars to sketch a square $ABCD$ with vertex A at the origin. Color the square. Select A and C, in that order, and apply **Transform | Mark Vector**. Then select the interior of the square and apply **Transform | Translate**. How could you tell Sketchpad to sketch also the boundaries and vertices of your translated square?

EXAMPLE 5

Continuing, place an independent point E in Quadrant I. Select, in order, B, A, E, and apply **Transform | Mark Angle**. Place an independent point F in Quadrant III. Select F and apply **Transform | Mark Center**. Next, select the square $ABCD$, including vertices, sides, and interior, and apply **Transform | Rotate**.

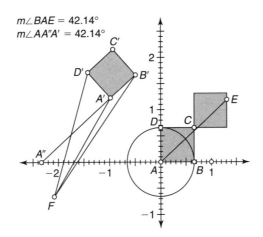

$m\angle BAE = 42.14°$
$m\angle AA''A' = 42.14°$

FIGURE B Translation and Rotation

EXAMPLE 6

Sketch a triangle ABC, including its interior, and place an independent point D outside $\triangle ABC$. Select D and apply **Transform | Mark Center**. Select all features of $\triangle ABC$, and apply **Transform | Dilate**. In the dialog box, accept the default ratio of $1/2$, but note that you could change it.

EXAMPLE 7

Continuing, sketch a line EF, and while it is selected, apply **Transform | Mark Mirror**. Then apply **Edit | Select All**, followed by **Transform | Reflect**.

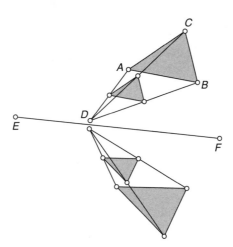

FIGURE C Dilation and Reflection

Shortcuts

As you know, there are two different ways to "go back" during a sketch. One is to apply **Edit | Undo**. In the process of using that command, when the **Construct** menu appears, to the right of **Undo**, you see **Ctrl Z**. That's the second way to "go back", and it is a good example of a shortcut. On the **Construct** menu, you can see several more shortcuts.

A combination that is particularly handy applies to the midpoint of two selected points. First, let's see how to do it using Windows: apply **Alt C | S** followed by **Ctrl M**. As you see, **Alt C** is a shortcut for **Construct**; similarly, you can work faster if you habitually use **Alt F** for **Files**, use **Alt E** for **Edit**, and so on. On Mac, there is no quick equivalent for this use of the **Alt** key. However, you can quickly sketch a midpoint of two selected points using ⌘**L** followed by ⌘**M**.

Another kind of shortcut is to perform certain constructions and measurements to several objects all at once, as illustrated in the next two examples.

EXAMPLE 8

Sketch a line and four points. Select all five objects and apply **Construct | Parallel Lines**. Click the screen to de-select all objects, and then select a line and the four points, and apply **Construct | Perpendicular Lines**. The latter, for example, can be done as **Alt C | d** (Windows) as indicated by "Perpendicular" in the **Construct** menu.

EXAMPLE 9

Sketch 8 independent points and apply **Edit | Select All**. Apply **Measure | Coordinates**. You should see all 16 coordinates instantly. To see them in action, select all the points again and apply **Animate | Points**. Note the continual updating of coordinates.

EXAMPLE 10

Suppose you wish to graph the equations:

$$y = \sin x + \cos x + \sin 2x + \cos 2x \quad \text{and} \quad y = \sin x + \cos x + \sin 2x - \cos 2x.$$

Start with **Graph | Define Coordinate System**, and apply **Graph | New Function**. In the resulting dialog box, you can type or click-in the first function and then apply **Graph | Plot Function**. For the second function, select Sketchpad's equation for the first function, and apply **Edit | Copy** followed by **Edit | Paste**. Drag the pasted copy to a convenient place, and, leaving it selected, apply **Edit | Edit Function**. This enables you to change the final + to −, thereby quickly formulating the second function while leaving the first one intact. Be sure to graph the second function.

As Example 10 suggests, Sketchpad has much to offer for the graphing of functions. With the two graphs still on the screen, select one of them and apply **Edit | Properties | Plot**. Then you can control the **Domain**, number of **Samples**, and whether to display a **Continuous** or **Discrete** graph.

SECTION 7   Custom Tools and Scripts

It is sometimes helpful to view a list of all the givens and steps that comprise a sketch. Such a list is called a **Script**. In order to view a Script, you must save or retrieve the sketch in the form of a **Custom Tool**. (Henceforth, we'll refer to a "custom tool" as simply a "tool".) For present purposes, let's imagine that you've completed a sketch and you wish to see its Script. Details will be given early in Chapter 2, but we may summarize the steps as follows: select the "givens" of your sketch and the results that you wish to be able to reproduce quickly in future sketches; create a tool for that purpose, and save it. Then, whenever you retrieve and apply the tool, you'll have the opportunity to apply **Custom Tool | Show Script View**. Keep in mind that **Script View** is not just for tools; rather, it is the way to see the Script of *any sketch*.

SECTION 8   Special Words

There are several words that have special meanings in connection with Sketchpad. Two of these are *parent* and *child*. Both occur, for example, in the **Edit** menu. Within any Sketchpad construction, an object U is a *parent* of object V if deleting U deletes V. Conversely, object M is a *child* of object N if N is a parent of M. For example, when you make a tool, every given is a parent of every result. When you select a point P and apply **Locus**, the locus is a child of P. When you place a movable point on a circle, the circle becomes a parent and the point, a child.

The classical meaning of *to construct* is *to define as a sequence of applications of the straightedge and compass in accord with the strict rules of Euclidean geometry*. In other words, a Euclidean construction is an ordered list of steps that are consistent with the Euclidean postulates. However, in many contexts, noneuclidean "steps" are allowed, and the results are still called constructions. On Sketchpad, **Measure | Calculate** enables many noneuclidean constructions.

In connection with Sketchpad, *sketch* is often used synonymously with *construction*. The word *sketch* has the appeal of being shorter, and, also, *sketch* is what *Sketch*pad does. Accordingly, a *sketch* may include captions and measurements.

There appears to be no term that distinguishes a construction from a "drawing that does not illustrate a geometric property". For our purposes, let an example suffice: to say "sketch a line perpendicular to a given line" means to use **Construct | Perpendicular Line**. You can't just drag a second line until it *looks* perpendicular to a given line, because, on further dragging, the lines will fail to stay perpendicular.

In this book, geometric objects are understood to be sets of points. One advantage of this understanding is that the point of intersection of lines

AB and *CD* can be written concisely as $AB \cap CD$. We may also write $V \cup W$, as when *V* and *W* are disks and we wish to sketch the centers of mass of *V*, *W*, and $V \cup W$ in order to confirm that these centers are collinear. A third contribution from set-theory is that line segments are infinite. Sometimes it is said that a segment is "finite", when what is meant is that it is *bounded*. The *length* of a segment is finite, but a segment and its length are two different things, as indicated by the notation *AB* for a segment and |*AB*| for its length. (Sketchpad, however, when printing lengths, leaves off the vertical marks.)

Sketchpad does a marvelous job of confirming geometric properties, often enabling the user to see that the properties hold in thousands of different cases, conveniently indexed by continuous motion. Such *confirming* may seem more convincing than *proving,* especially in the case of long and difficult proofs. But sketching is inherently different from proving. A proof is a sequence of propositions, starting (if all the steps are included) with those basic special ones called postulates and ending with one called the conclusion. To sketch is to visualize, whereas to prove is to deduce.

SECTION 9 Documents and Printing

You are probably familiar with most of the items in the **File** menu because they are common to many software programs. The two items that will be considered here are **Document Options** and **Print Preview**. The kind of file saved by Sketchpad is called a Document. A Document can consist of sketches of two kinds: pages and tools. It is quite common for a Document to contain one page and nothing else; the page is simply a sketch. However, if you have several sketches that belong together, they can be the pages of a Document. When the Document is opened, all its pages and tools will become quickly accessible using **Document Options**.

Before printing a sketch, be sure to apply **File | Print Preview**. The button **Fit To Page** enables you to automatically increase or decrease the scale. You can also control the scale by typing in a desired percentage.

For some sketches, lines that extend from your configuration may take up too much space on paper, and the part you really care about might be too small. This is a common problem, because for many geometric purposes, it is better to use lines than segments. (For example, sketch the altitude from *A* onto segment *BC* in an acute $\triangle ABC$, and then drag *A* so that $\angle ABC > 90°$.) In order to solve this problem, you can *crop* the lines: on each extended line place a pair of points *U* and *V* whose segment you want to keep; construct the segment, then de-select it, select the line, and apply **Display | Hide** (or **Alt D|H**) (Windows). You can hide the points *U* and *V*, too.

A final word on the subject of printing concerns the printing of Scripts. This is possible when the **Script View** box is visible: right-click the mouse on any object listed in the box, and then click on **Print**.

CHAPTER ONE

Simple Sketches

ABOUT 2300 YEARS ago, Euclid summarized what was known about geometry in *The Elements,* said to be the second-most widely studied book ever written. Since Euclidean geometry is perceived as an expression of pure reason, it is one of the oldest and most enduring of all subjects. In particular, this subject has, for centuries, held a high place in school curricula because of its applicability to the development of the ability to think.

During most of the two and a half millenia since the infancy of geometry, sketches were done by hand, using the Euclidean tools—a straightedge and compass. Now we have computer-based sketching. What's more, the computer can "move" your sketches, so that you can observe how various geometric relationships among the objects change or remain unchanged. It is this movability that gives this book its name, *Geometry in Action.*

The Geometer's Sketchpad lets you carry out virtually all geometric constructions that follow strict Euclidean rules. However, Sketchpad's ability to go *beyond* Euclidean steps opens up vast new worlds of possibilities. You can plot ornate curves using **Locus**, and trisect angles using **Measure | Calculate**. You can graph elaborate functions (Chapter 5), and explore noneuclidean geometry (Chapter 9).

The main objective here in Chapter 1, however, is to take a good first step into the field of Euclidean geometry. We'll start out with two of the most basic notions of geometry (indeed, two of the most basic notions of human thought): points and lines.

Points and Lines

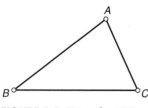

FIGURE 1.1 Triangle *ABC*

The most common "givens" that determine a line are two points. On Sketchpad, after putting two points on the screen, you can easily construct their line, or, just as easily, the segment between the two points. If you start with three points, *A*, *B*, *C*, not already in a line, you can quickly construct from them three segments. This construction includes the triangle, △*ABC*, as in Figure 1.1.

Centroid

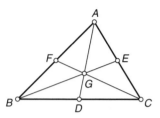

FIGURE 1.2 Triangle *ABC* and its centroid, *G*

If you then sketch segments from *A* to the midpoint of *BC*, from *B* to the midpoint of *CA*, and from *C* to the midpoint of *AB*, you'll have Figure 1.2.

A theorem proved in many geometry books is that these particular three segments meet in a point. Traditionally labeled *G*, as in gravity, this point is known by the names *center of gravity, center of mass,* and *centroid*.

The segments *AD*, *BE*, *CF* are called *medians*. They afford a good opportunity to use Sketchpad's **Measure** and **Calculate** capabilities. To measure the length of segment *AD*, select *A* and *D*, and apply **Measure | Distance**. You'll see the length printed on the screen. If you move anything that changes this length, the measurement will be instantly updated. Another way to see the same measurement is to have *AD* constructed as a segment, and to apply **Measure | Length**. You can change the units and precision for measurements using **Edit | Preferences**.

To exemplify the use of **Calculate**, let's calculate the ratio $|AD|/|GD|$. First, measure both $|AD|$ and $|GD|$. Then select both. Applying **Measure | Calculate** then causes a **Calculator** to appear. On it, click **Values**, and you'll see the two measurements printed. Click the first, then click the division key (÷), then click **Values**, then the second measurement, and click **OK**. The desired ratio will appear.

How to Name Your Sketches

Just below, you see Assignment 1.1, consisting of sketches for you to create and possibly submit for evaluation. In this book, every Assignment is essentially a list of problems, with names like **SS1A** and **SS1B**. *Use the given name whenever you save a sketch.* Also, put the name in a caption and tell briefly what a user should know when running your sketch. For example, your first sketch should be saved with filename **SS1A** and caption like this: **SS1A. Triangle ABC with medians meeting in centroid G. Drag points A, B, C.**

ASSIGNMENT 1.1

SS1A. Sketch a triangle *ABC*, its medians *AD*, *BE*, *CF*, and centroid *G*. Print a caption that starts with the name **SS1A** and tells briefly what geometric proposition **SS1A** illustrates. Include instructions for running the sketch. Save your sketch as **SS1A** (which is short for **S**imple **S**ketches, Section **1**, Sketch **A**).

SS1B. Add to **SS1A** measurements of the areas of triangles *BGC*, *CGA*, *AGB*.

SS1C. After saving **SS1B**, add to it an independent point *P*. Then measure the areas of triangles *BPC*, *CPA*, *APB*. Vary *P* so as to discover all positions of *P* for which the three triangles have equal areas.

SS1D. Starting with **SS1A**, add a calculation of the ratio, $|AG|/|DG|$. Then add an independent point *P*, and sketch the lines *AP*, *BP*, *CP* and the points

$$Q = AP \cap BC \quad R = BP \cap CA \quad S = CP \cap AB$$

Vary *P* to find out if the ratio $|AP|/|QP|$ equals 2 only when $P = G$. (Remember, always print your conclusions in a caption.)

Circumcenter

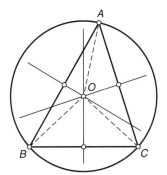

FIGURE 1.3 Triangle *ABC* and its circumcenter, *O*. $|AO| = |BO| = |CO|$ = radius of circumcenter

The objective here is to construct a point that is equidistant from three given points. The three points, *A*, *B*, *C*, form a triangle, and the point to be constructed is called the *circumcenter* of △*ABC*, as in Figure 1.3.

To start, there is a point that is clearly equidistant from two of the vertices, *B* and *C*, namely, their midpoint. Are there some other points, near that midpoint, that are also equidistant from *B* and *C*? Feel free to experiment using a pencil and paper—but don't read beyond this sentence until you have convinced yourself that there are many points equidistant from *B* and *C*.

The set of *all* points equidistant from *B* and *C* is a line; it is perpendicular to *BC* and passes through the midpoint of *BC*. While all the points on this line have equal distances from *B* and *C*, their distances from *A* vary. One of these points, we shall see, has just the right distance from *A*, too.

Again, pause and try to discover, with pencil and paper, or with Sketchpad, some way to determine the special point whose distance from *A* equals that from *B* and from *C*. Again, it's best to discover this on your own before reading further.

Here's one way to do it: since the points on the perpendicular bisector of *BC* are equidistant from *B* and *C*, and since those on the perpendicular bisector of *CA* are equidistant from *C* and *A*, a point on both bisectors must be equidistant from all three points. The two bisectors are shown in Figure 1.3.

ASSIGNMENT 1.2

SS1E. Sketch △*ABC*, the perpendicular bisectors of the sides, and their point of concurrence. Label the point *O*, this being the traditional symbol for the circumcenter.

SS1F. After saving **SS1E**, add segments *AO*, *BO*, *CO* and measurements of their lengths. Use **Measure | Calculate** to confirm that this common length, called the *circumradius,* equals

$$\frac{|BC| \cdot |CA| \cdot |AB|}{4D}$$

where *D* denotes the area of △*ABC*.

SS1G. Starting with **SS1E**, place an independent point P inside $\triangle ABC$, and convince yourself that the lengths $|AP|$, $|BP|$, $|CP|$ are equal only if $P = O$. Then drag P to the outside of $\triangle ABC$ and find as many other positions as you can for which the three lengths are equal.

Orthocenter

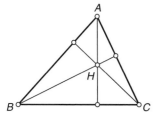

FIGURE 1.4 Triangle ABC and its orthocenter, H

For a triangle ABC, the word *altitude* has two meanings: the distance from a vertex to the opposite side, and the line (or segment) through a vertex, perpendicular to the opposite side. Using this second meaning, we can say, as did the ancient Greeks, that the three altitudes concur in a point. The point where the three altitudes meet is the *orthocenter,* with traditional symbol H, as shown in Figure 1.4.

ASSIGNMENT 1.3

SS1H. Sketch $\triangle ABC$, the altitudes, and H. Predict what will happen when you drag each of the points A, B, C, H. Then use your sketch to confirm or refute your predictions.

SS1I. After saving **SS1H**, construct the line through A parallel to BC, the line through B parallel to CA, and the line through C parallel to AB. Then construct the circumcenter of the triangle formed by those three lines. Try to measure the distance between this point and H. (As always, if you discover something noteworthy, describe it in a caption.)

Euler line

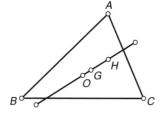

FIGURE 1.5 Triangle ABC and its Euler line

So far, we've constructed three centers of a triangle: centroid (G), circumcenter (O), and orthocenter (H). A popular problem is to prove that these points are collinear; that is, that they lie in a line. This line is named the *Euler line* after Leonhard Euler (pronounced oil´er); you can find several websites about Euler and his prolific contributions to mathematics.

ASSIGNMENT 1.4

SS1J. Sketch $\triangle ABC$ and its Euler line. By dragging vertex A, determine triangles for which the Euler line passes through a vertex.

SS1K. After saving **SS1J**, use **Measure | Calculate** to display the ratio $|GH|/|GO|$. Does this ratio vary as you change the shape of $\triangle ABC$?

SECTION 2   Circles

In this section, we'll start with applications of Sketchpad's two **Construct | Circle** commands. First, we'll construct an equilateral triangle using a given segment as one of the sides, followed by constructions named for Fermat and Napoleon. (Pierre Fermat made his living as a lawyer but is ranked among the greatest mathematicians; Napoleon Bonaparte was Emperor of the French and knew several leading mathematicians.)

Equilateral Triangles and Fermat Point

A circle with center O and radius r will be denoted by $\circ(O, r)$. Consider Figure 1.6, where $\circ(B, |BC|)$ meets $\circ(C, |BC|)$ in two points, labeled D and D'.

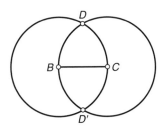

FIGURE 1.6 Points D and D' equidistant from points B and C

Since D is on $\circ(B, |BC|)$, its distance to B is $|BC|$, and since D is on $\circ(C, |BC|)$, its distance to C is $|BC|$. In other words, $|DB| = |DC| = |BC|$, which is to say that $\triangle BCD$ is equilateral. Consequently, the three angles are equal, and since their sum is $180°$, each has a measure of $60°$.

In the history of mathematics, four triangle centers were known to the ancient Greeks, and the next one came along many centuries later, when Fermat posed the following problem: *in the plane of $\triangle ABC$, find a point P for which the sum*

$$|PA| + |PB| + |PC|$$

of distances is as small as possible. The solution was eventually published in 1659, and the point P is often called the *Fermat point*.

To construct the Fermat point of $\triangle ABC$, sketch an equilateral triangle $A'BC$ with A' on the side of line BC that does not contain vertex A. Likewise, sketch equilateral triangles $B'CA$ and $C'AB$. Then construct lines AA', BB', CC'. They concur in the Fermat point.

ASSIGNMENT 2.1

SS2A. Construct $\triangle ABC$ and its Fermat point, labeled F.

SS2B. Continuing, add an independent point P, and apply **Measure | Calculate** to evaluate the sum

$$|PA| + |PB| + |PC|$$

Vary P to determine as many points as you can for which this sum attains its least possible value.

SS2C. Starting with **SS2A**, add an independent point P and measure $\angle BPC$, $\angle CPA$, and $\angle APB$. Vary P to determine as many points as you can for which these angles are equal.

Napoleon Points and Napoleon Triangle

Figure 1.7 shows $\triangle ABC$ with equilateral triangles based on the sides BC, CA, AB. The construction of the Fermat point uses the outer vertices of the equilateral triangles. Here, instead, we'll use the centroids of these triangles.

The *Napoleon Theorem* states the fact that $\triangle UVW$, as shown in Figure 1.7, is equilateral. The triangle itself is called the *Napoleon triangle*. Lines AU, BV, CW concur in the *Napoleon point*. It is not known whether Napoleon discovered the theorem, triangle, and point that bear his name.

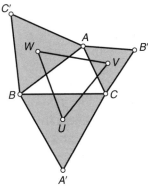

FIGURE 1.7 Centers U, V, W of equilateral triangles

SS2D. Construct △*ABC* and its Napoleon triangle, *UVW*. Print measurements of ∠*UVW*, ∠*VWU*, ∠*WUV*. If they aren't 60°, something is amiss.

SS2E. After saving **SS2D**, add lines *AU, BV, CW* concurring in the Napoleon point, labeled *N*.

SS2F. A glance at Figure 1.6 suggests that there are *two* equilateral triangles based on a segment. Up to now, we've used those directed *outward* from △*ABC*. By reflecting *A′* in *BC*, you get the vertex *A″* of an *inward* equilateral triangle. Similarly, obtain *B″* and *C″*, so that you have three new equilateral triangles. Construct the centroids *U, V, W* of the new equilateral triangles, and discover whether the lines *AU, BV, CW* concur.

Circumcircle

Two points determine a line, and three noncollinear points determine a circle. One of the most common Euclidean constructions is that of the circle that passes through three such points. They form a triangle, of which the circle is called the *circumcircle.*

Suppose *ABC* is a triangle. If we can find a point whose distances from *A*, *B*, and *C* are equal, then this common distance must be the radius of the circumcircle, as in **SS1F**.

SS2G. Construct △*ABC* and its circumcircle.

SS2H. Use **Construct | Arc Through 3 Points** to sketch an incomplete circle. Figure out how to sketch its center, then sketch it.

Incircle and Excircles

The circumcircle deserves its name because it *circum*scribes the given triangle; another important circle is the *incircle,* and it is *in*scribed, as shown in Figure 1.8.

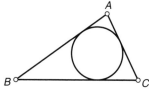

FIGURE 1.8 △*ABC* and its inscribed circle

To construct the incircle of △*ABC*, we need the center of the circle and a pass-through point. This center is the fourth ancient Greek point, in addition to *G*, *O*, and *H*; it is the *incenter,* traditionally denoted by *I*.

To construct it, start with this question: where are the points *P* that have equal distances from sides *AB* and *AC*? Certainly one such point is *A*, for which these distances are zero. It would be a good idea to experiment with pencil and paper, or Sketchpad, trying to find other points *P* for which the two distances (measured from *P* perpendicularly to the lines) are equal. When you've found several such points, go on to the next paragraph.

The required points form the bisector of ∠*CAB*. Use Sketchpad to place a movable point *P* on the bisector, to construct perpendiculars from *P* to the sidelines *AB* and *AC*, and to measure the two perpendicular distances. Then move *P* on the bisector and note that the two distances stay equal. (If *P* were not on the bisector, which distance would be the smaller?)

If you bisect ∠*ABC* also, then this new bisector *must* meet the bisector of ∠*CAB*. Why? Being on the old bisector, the point of intersection is equidistant from lines *AB* and *AC*, and being on the new bisector, it is equidistant from lines *AB* and *BC*. So, it is equidistant from all three sidelines of △*ABC*.

Actually, there are more points that are equidistant from the sidelines. One, called the *A-excenter,* is the point of intersection of the (internal) bisector at vertex *A* and the external bisector at vertex *B*. The external bisector at vertex *C* also goes through the *A*-excenter.

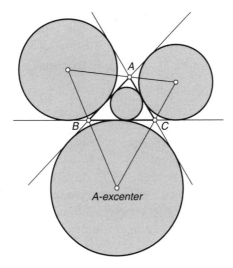

FIGURE 1.9 △*ABC* and its incircle and excircles

Let's be sure about "external bisector". For example, let *A′* be the reflection of *A* about *B*. That is, on Sketchpad, select *B* and apply **Transform | Mark Center**, then select *A* and apply **Transform | Rotate** using 180°. The external bisector at vertex *B* is now explicitly defined as the bisector of ∠*CBA′*.

ASSIGNMENT 2.4

SS2I. Construct △*ABC* and its incircle.

SS2J. Continuing, extend the sidelines of △*ABC* (if they were formerly segments). Then add the three excircles. Print a measurement of the angle between the interior and exterior bisectors at point *A*.

SS2K. Continuing, add measurements of the radii of the four circles. Let r denote the inradius, and s, t, u the exradii. Use **Measure | Calculate** to print measurements of $1/r$ and $1/s + 1/t + 1/u$. As you vary the shape of △*ABC*, how are these two measurements related to each other?

SECTION 3 # Constructions Using Measurements

As already noted in Section 1, Sketchpad enables reasonably accurate constructions that are not possible under the strict Euclidean rules using

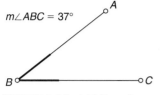

$m\angle ABC = 37°$

FIGURE 1.10 $\angle ABC$, ready for trisection

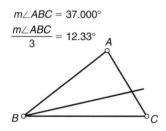

$m\angle ABC = 37.000°$

$\dfrac{m\angle ABC}{3} = 12.33°$

FIGURE 1.11 $\angle ABC$ trisected

straightedge and compass. Probably the most often cited example of such a construction is angle-trisection.

1. Measure $\angle ABC$. (When rotating by an angle, the direction is important, so to measure this angle, select the points in the order A, B, C.)

2. Select the printed measurement of $\angle ABC$ that appears on the screen.

3. Next, choose **Measure | Calculate**. In the resulting dialog box, click **Values**, and then click on the printed measure of $\angle ABC$. Click on ÷ and type **3** (or click **3**), and click **OK**.

4. Select point B, and apply **Transform | Mark Center**. Select the printed measurement of the one-third angle. Then apply **Transform | Mark Angle**.

5. Select segment BC, and apply **Transform | Rotate by Marked Angle**.

Measurements of your angles will be reasonably accurate if you use **Edit | Preferences** to set **Angle Precision** to thousandths; this may be done before you start sketching, or afterwards if you apply **Edit | Select All** before applying **Edit | Preferences**. By experimenting with your trisected angle, you'll find that the original angle must measure less than 180°.

A hundred years ago, high school geometry teachers encouraged students to try to trisect an arbitrary angle using the strict Euclidean rules. (Meanwhile, science teachers encouraged students to try to devise a perpetual motion machine). Thousands of students took up the challenge, and many kept at it their whole lives. None succeeded. However, so many people have tried that university mathematics departments have received thousands of attempts. The trisection craze and many of the cleverest attempts are documented in Underwood Dudley's *A Budget of Trisections,* (Springer-Verlag, 1987).

Morley Theorem and Morley Triangle

One of the most famous theorems of early twentieth-century geometry was discovered by Frank Morley. This theorem is of particular interest because it involves trisected angles.

In a triangle ABC, there are two angle trisectors through each vertex. Starting at vertex B, we have one trisector that is closer to side BC, and the other, closer to side BA. In Figure 1.12, these are labeled bc and ba. Likewise, the other trisectors are labeled ca and cb, and ab and ac.

Let A' be the point in which lines bc and cb meet; similarly, $B' = ca \cap ac$ and $C' = ab \cap ba$. The triangle $A'B'C'$ is called the Morley Triangle, and Morley's Theorem states that this triangle is equilateral.

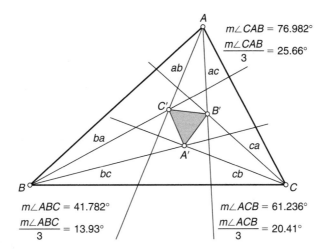

FIGURE 1.12 Triangle *ABC*, its six internal angle trisectors, and the Morley triangle

$m\angle CAB = 76.982°$

$\dfrac{m\angle CAB}{3} = 25.66°$

$m\angle ABC = 41.782°$

$\dfrac{m\angle ABC}{3} = 13.93°$

$m\angle ACB = 61.236°$

$\dfrac{m\angle ACB}{3} = 20.41°$

ASSIGNMENT 3.1

SS3A. Starting with △*ABC*, construct its Morley Triangle, and print measurements of

$$\angle B'A'C' \quad \angle C'B'A' \quad \angle A'C'B'$$

Be sure they stay equal to 60° when you drag *A*, *B*, *C*.

SS3B. After saving **SS3A**, add lines *AA′*, *BB′*, *CC′*. Do you think these lines concur, regardless of the shape of △*ABC*?

SS3C. The three great problems of antiquity were to find Euclidean constructions for trisecting an angle, "squaring a circle", and "duplicating a cube". (There are algebraic proofs that these constructions are not possible using only the Euclidean tools.) Use **Measure | Calculate** to perform a Sketchpad construction for the second great problem; that is, starting with an arbitrary circle, sketch a square having the same area.

Analytic Geometry: Centroid of a Set of Points

First, what is meant by "analytic" geometry? The answer is simple: geometry that uses coordinates. Sketchpad enables all sorts of good things using coordinates. One venture that introduces many of these capabilities is to sketch the centroid of a set of points. This was already done in Section 1 for a set of three points. Now we'll start with four points. The centroid, often denoted by $(\bar{x}, \bar{y})$ is given by

$$\bar{x} = (x_1 + x_2 + x_3 + x_4)/4 = \text{average of the } x \text{ coordinates of the points}$$

$$\bar{y} = (y_1 + y_2 + y_3 + y_4)/4 = \text{average of the } y \text{ coordinates of the points}$$

To realize these averages in a sketch, start with four points, labeled *A*, *B*, *C*, *D*. Select all of them and apply **Construct | Segment**. Then drag vertices, if necessary, so that they form a quadrilateral.

1. Apply **Graph | Define Coordinate System**. Select *A*, *B*, *C*, *D* and apply **Measure | Abscissa**. Then apply **Measure | Calculate** to compute $\bar{x}$; similarly, compute $\bar{y}$.

2. Select the printed values for $\bar{x}$ and $\bar{y}$, in that order, and apply **Graph | Plot As** (x, y).

3. Be sure to move A, B, C, D and watch their centroid "follow". Also, examine the following variations:
   - Drag the origin and watch all the printed coordinates change.
   - Drag the point at 1 on the x-axis to rescale the axes; again, coordinates follow.
   - Drag point A to a position for which the centroid lies outside the polygon.

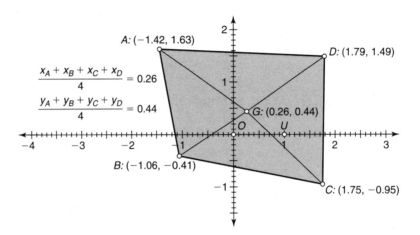

FIGURE 1.13 Centroid, G, of four points, A, B, C, D

Now a word about being classy: you work hard on your sketches, so they deserve some class. Following are some suggestions, but the important thing is to suit yourself according to your preferences and tastes: fill polygon-interiors with light colors; use thick boundary lines; drag Sketchpad's printed measurements into coherent clusters; in your captions; use bold-face print for emphasis; maybe use a distinctive font. Choose labels that match geometric relationships. Adjust the positions of labels so that they are not obscured. Word your captions so clearly that they cannot reasonably be misunderstood.

ASSIGNMENT 3.2 **SS3D.** Follow the steps printed above for sketching the centroid of a set of four independent points.

SS3E. Starting with five or six independent points, sketch their centroid.

PROJECTS

Each chapter in this book ends with a selection of Projects. These lend themselves well to working in groups of size one to four. If you are in a position to choose which projects to work on, you should first scan all of them, as there are considerable ranges of topics and levels of difficulty. The instructions are intentionally concise in order to promote creativity. Feel free to explore mathematical possibilities and to consult **Help** or the online Geometer's Sketchpad Resource Center at **http://www.keypress.com/sketchpad/**.

How to Name Project Sketches

The filename, when you save your sketch for Project 1, should be **Ch1Pr1**. For the parts of Project 2, use names **Ch1Pr1**, **Ch1Pr2Part1**, **Ch1Pr2Part2**, etc. You'll find that long, explicit names are very helpful when managing your files.

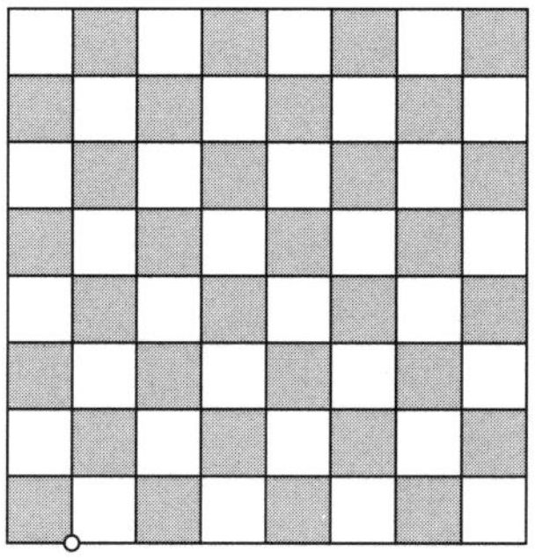

PROJECT 1: CHESSBOARD

Using as few Sketchpad steps as possible, emulate Figure 1.14. Be sure to use options in the **Transform** menu. The objective here is not just to obtain the right image, but to do so in as few steps as you can.

FIGURE 1.14 Chessboard

PROJECT 2: GEOMETRIC ARITHMETIC

This project introduces the most fundamental connections between geometry and algebra. It is conceptually very interesting that the ancient Greeks did not have "algebra". What we call x^2, for example, was a geometric square with sidelength x. Similarly, the numbers that we so conveniently write as

$$\sqrt{x} \quad x+y \quad x-y \quad x/2$$

were regarded and represented geometrically.

Part 1. Start with arbitrary segments. Let x and y denote their lengths. Figure out how to construct a circle of radius $(x + y)/2$. (Throughout Project 2, you should be able to vary the lengths of initial segments. In this project, "to construct" means "to construct in the Euclidean manner"; that is, without using **Measure**)

Part 2. Start with two segments of arbitrary length. Decree the length of one of them to be 1 unit, and let x denote the length of the other. Figure out how to construct a segment of length $1/x$. Hint: similar triangles ABC, ABD, ADC in Figure 1.15.

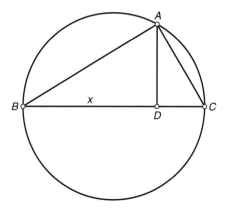

FIGURE 1.15 Hint for constructing a segment of length $1/x$

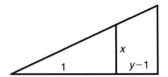

FIGURE 1.16 Hint for constructing a segment of length xy

Part 3. Start with segments of lengths 1, x, y. Figure out how to construct segments of lengths xy, x/y, and y/x. Hint: similar triangles in Figure 1.16.

Part 4. Start with two segments of arbitrary length. Decree the length of one of them to be 1 unit, and let x denote the length of the other. Figure out how to construct a segment of length $\sqrt{x}$. (Hint: Figure 1.15) Then for arbitrary length y, figure out how to construct a segment of length $\sqrt{xy}$.

An excellent reference in connection with Project 2 is Howard Eves, *An Introduction to the History of Mathematics, 6th edition,* Saunders, New York, 1990, especially pages 85-90.

PROJECT 3: SLIDERS AND TRIANGLES

A slider is a segment whose length can vary. Many Sketchpad constructions use one or several sliders. When there is more than one, it is a good idea to arrange them like the bars of a bar graph.

Part 1. Construct three sliders, with labels a, b, c for their lengths. For some choices of a, b, c, there exists a triangle having sidelengths a, b, c. Figure out conditions for a, b, c for which such a triangle exists, and sketch it. Label your triangle ABC, and print measurements of a, b, c, A, B, C. (Here, as is common, the symbols A, B, C denote vertex angles as well as vertex points.)

Part 2. Start with three independent points A, B, C. Print the lengths u, v, w of the medians of $\triangle ABC$. Construct a triangle UVW whose sidelengths are u, v, w, as in Figure 1.17. For what vertex angles A, B, C will $\triangle UVW$ be similar to $\triangle ABC$?

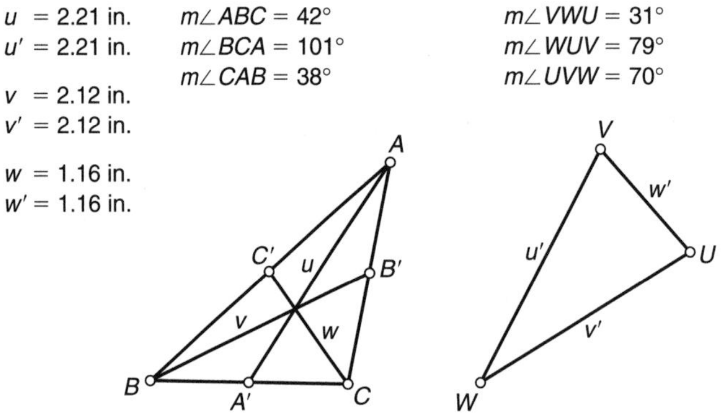

$u = 2.21$ in. $m\angle ABC = 42°$ $m\angle VWU = 31°$
$u' = 2.21$ in. $m\angle BCA = 101°$ $m\angle WUV = 79°$
         $m\angle CAB = 38°$ $m\angle UVW = 70°$
$v = 2.12$ in.
$v' = 2.12$ in.

$w = 1.16$ in.
$w' = 1.16$ in.

FIGURE 1.17 Can triangles ABC and UVW be similar?

PROJECT 4: COUNTING

Part 1. Start with four points, no three of which are collinear. Join every pair of your points by a line. How many such lines are there? Repeat the experiment using five points. Be sure to vary the points to see that such variations can change the number of lines determined. Repeat using six points. For arbitrary $n \geq 3$, figure out how many lines are determined by n points, assuming that no two of them are collinear. See Figure 1.18.

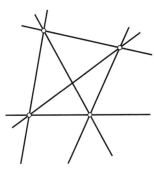

FIGURE 1.18 Lines determined by four points

Part 2. Start with four lines, no two of which are parallel and no three of which concur. (The word *concur* is used when three or more lines meet in a single point.) Each pair of lines meet in a point. How many such points are there? Repeat the experiment using five lines. Be sure to vary the lines to see that such variations can change the number of points determined by the five lines. Repeat using six lines. For arbitrary $n \geq 3$, figure out how many points are determined by n lines, assuming that no two are parallel and no three concur.

PROJECT 5: LENGTH AND AREA

Sketch independent points A, B, C and their triangle.

Part 1. Let R and r denote the circumradius and inradius of $\triangle ABC$. Use **Measure | Calculate** to sketch a segment of length R/r, where $R =$ circumradius and $r =$ inradius. Figure out with pencil and paper the value of R/r when $\triangle ABC$ is equilateral, and then use your sketch to confirm your figuring.

Part 2. Continuing, let

$$s = (|BC| + |CA| + |AB|)/2$$

Print in a caption a proof that

$$\text{area}(\triangle ABC) = rs$$

and use **Measure | Calculate** to confirm this formula.

Part 3. Without using **Measure | Calculate**, sketch a segment whose length equals the area of $\triangle ABC$. Hint: see Project 2.

PROJECT 6: AUBEL'S THEOREM

Part 1. Sketch a quadrilateral using independent points as vertices. Sketch a square on each of the four sides, facing outwards. Join the centers of opposite squares with segments. Apply **Measure | Calculate** to confirm that these two segments have equal lengths and are perpendicular, as shown in Figure 1.19.

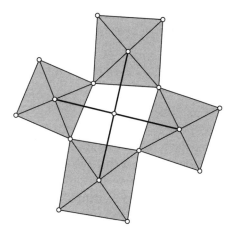

FIGURE 1.19 Perpendicular congruent segments

Part 2. Reverse the construction in **Part 1**. That is, create a new sketch, starting with perpendicular segments of equal lengths, and then construct the four squares centered and touching as in Figure 1.19. Be sure that these properties persist when you vary your initial segments.

PROJECT 7: CENTER OF MASS

Figure 1.20 shows a 3-dimensional tabletop. Strings attached to point G run through imaginary pulleys at the corners of the tabletop down to masses m_1, m_2, m_3, m_4. If, for example, mass m_1 increases, then point G, the center of mass (if the masses were on the tabletop), moves toward the corner above mass m_1.

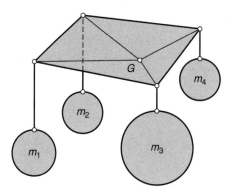

FIGURE 1.20 Tabletop and center of mass, G

In Figure 1.13, the same sort of configuration is shown, except that all the masses are equal to 1. In order to generalize to masses m_A, m_B, m_C, m_D located at corners A, B, C, D, respectively, the following formulas may be used to calculate the coordinates $\bar{x}$ and $\bar{y}$ of the center of mass, G:

$$\bar{x} = \frac{m_A x_A + m_B x_B + m_C x_C + m_D x_D}{m_A + m_B + m_C + m_D}$$

$$\bar{y} = \frac{m_A y_A + m_B y_B + m_C y_C + m_D y_D}{m_A + m_B + m_C + m_D}$$

These quotients are called weighted averages. When all four masses are equal, these formulas yield ordinary averages.

Part 1. Emulate Figure 1.13, but include sliders for your 4 masses. Include a printing of the masses (i.e., lengths of the sliders), and use the above formulas to sketch the centroid.

Part 2. Create a sketch like that for **Part 1**, representing 3 masses instead of 4.

Part 3. Create a sketch of a triangle T, a square S, and a circle C, together with their centroid. Assume that the masses of T, S, C are equal, and enable the user to drag the vertices of T and S, as well as the center and a pass-through point of C. (Hint: in the formulas for $\bar{x}$ and $\bar{y}$, use 3 coordinates instead of 4, and use areas for masses.)

CHAPTER TWO

Tools

TOOLS ARE SKETCHES ready for rapid deployment in longer sketches. Let's say you need to construct the centroids of four different triangles. You could repeat all the steps in **SS1A** (Figure 1.2) four times. Or, you could make a tool from **SS1A** and apply it four times. That would be much more efficient.

It will pay to build up a collection of tools. Like tools from a hardware store, some will be used over and over again. If you are in a position to save duplicates of your tools on a hard drive, as well as a portable disk, that would be wise. This is especially true if you have exclusive access to the folder in which the Sketchpad program resides, since in that case you can save tools in the special **Tool Folder**.

Actually, to "save a tool" is a short way to say "save a document that contains the tool". Once you've been through the process step-by-step, as in Section 1, it will be clear how tools are saved and retrieved.

S E C T I O N 1 ## Tools for Special Points

In Chapter 1, special points—the four ancient Greek triangle centers—were constructed: centroid, circumcenter, orthocenter, and incenter. Here, we'll see how to convert the construction of the centroid into a tool. Following this example, you can create other tools.

The Ancient Triangle Centers

Creating a tool consists of three steps: (1) open or create a sketch; (2) select the "givens" and "results" in the sketch; (3) apply **Custom Tool | Create New Tool**. Let's use **SS1A** (Figure 1.2) as an example.

1. Open **SS1A.gsp** (or start from scratch and construct the centroid G of a triangle ABC).

2. Select the vertices of the triangle. They may be selected in any order. Their labels, whether visible or hidden, need not be A, B, C. Then select G. It is necessary that the givens, namely the three vertices, be selected before the results, namely the centroid.

3. Click the **Custom Tools** icon. You'll see a dialog box with **Create New Tool** at the top. Click it. If your selections are not just right, you'll be so advised and can go back and try again. Otherwise, you'll see a dialog box into which you can type the name **centroid**. Click the button for **Show Script View.**

4. You now have a box headed **centroid Script**. It invites you to "Type your comments here". To accept this invitation, first click **OK** in the **New Tool** box. Then type "This tool sketches the centroid of the triangle of 3 selected points". (You can enlarge the window where you're typing by dragging its lower edge down.) Then close **Script View** by clicking the button in the upper right corner.

5. Save your work as **centroid** as you would ordinarily save a file, using **File | Save As**. Sketchpad automatically appends **.gsp**, so that the full name of your document is **centroid.gsp**, and, saved as a part of the document is the tool named **centroid**. To confirm this, apply **File | Document Options**, and click **Tools**.

A Sketchpad document can contain several pages and several tools. In this book, however, the documents to be discussed will consist of single pages, together with one or zero tools. In the future, we'll write a name like **centroid** for *both* the document and the tool, and you'll be able to tell which is intended by context, such as "Open **centroid**" or "Apply **centroid**". To "save a tool" means to save the document that contains the tool.

Here are some suggestions regarding tools.

- Be an avid tool collector. Every time you create a worthwhile sketch, use it to create a tool. When you save the sketch, the tool will be saved with it.

- If you have access to the folder containing the Sketchpad program, then you may wish to save tools in the **Tool Folder**. These tools will automatically become ready-to-open whenever Sketchpad is opened. If someone else has access to the **Tool Folder**, you should build up your own main tool collection on a separate disk. You should also maintain a backup of your tool collection—you'll be glad you did!

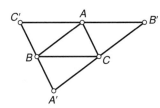

FIGURE 2.1 Triangle *ABC* and its anticomplementary triangle *A'B'C'*

Next, let's apply **centroid** to several triangles in a sketch. Start with △*ABC*, and add the parallel to *BC* through *A*, the parallel to *CA* through *B*, and the parallel to *AB* through *C*, as shown in Figure 2.1. The three parallels form a triangle *A'B'C'* long ago dubbed the *anticomplementary triangle*.

Now **centroid** can be applied to the points *A'*, *B*, *C* to construct the centroid of △*A'BC*, and then applied to the points *A*, *B'*, *C*, and to the points *A*, *B*, *C'*. Note that this tool, and all tools, perform without depending on particular labels. Even though the "givens" may have been labeled *A*, *B*, *C* in the creation of **centroid**, the tool applies to *any* three selected points, regardless of their labels.

Let's step through the application of the tool **centroid** to the points *A'*, *B*, *C*. The sketch in Figure 2.1, or its equivalent, must be active. Open **centroid.gsp** if it is not already open. Click the **Custom Tool** button in Sketchpad's main toolbox. You'll see a dialog box, including the name of centroid as a document. Place the mouse's arrow on the name centroid, and notice that another box appears, also labeled **centroid**; click it. (This second one is the tool; the first is the document that contains the tool.) Click the **Custom Tool** button again, and near the top of the box, click **Show Script View** to see a **Script box**.

At the top appears **centroid** followed by the word **Script**, indicating all the steps, in words and symbols. To view the whole script, you can use the two scroll bars in the dialog box, or, better, you can make a printout on paper by right-clicking on the interior of the dialog box and then clicking on **Print**.

The interior of the dialog box consists of sections headed **Given:** and **Steps:**. At this point, de-select any objects on your screen that may have been selected, and select, in order, points *A'*, *B*, *C*. Watch the **Script box** when making selections, noting that they are recorded in the box. As soon as you've finished, the box will indicate two opportunities after the word **Apply**. Click the second one, labeled **All Steps**. You'll see a quick construction, after which its objects, except the final "results", will be hidden.

ASSIGNMENT 1.1

TO1A. Open **SS1A** and create and save a tool named **centroid**. Then start a new sketch, consisting of △*ABC* and its anticomplementary triangle *A'B'C'*, as in Figure 2.1. Apply **centroid** to sketch the centroids of triangles *BCA'*, *CAB'*, *ABC'*. Label these centroids *D*, *E*, *F*, and sketch the centroid of △*DEF*. Save the result as **TO1A**. (One last reminder: Sketchpad will add the **.gsp** and save your work as a one-page document; if you also want your work saved as a tool, it's up to you to create the tool, to name it, and to remember the name of the document that contains it, or else save the tool in another document that has the same name as the tool. The second option is recommended for tools that you expect to use often.)

TO1B. Use **SS1E** as a basis for creating a tool named **circumcenter** which constructs the circumcenter of a triangle. After saving this tool, open **TO1A**. Delete points *D*, *E*, *F*. Construct the circumcenters of triangles *ABC* and *A'B'C'* and the circumcircles of these triangles. Save the result as **TO1B**. Your tool, **circumcenter**, now resides in **TO1B**; do you want to save it also in a document named **circumcenter.gsp**?

TO1C. Use **SS1H** as a basis for creating **orthocenter**, which constructs the orthocenter of a triangle. Save **orthocenter**. Start a new sketch by constructing a triangle *ABC*. Apply the three tools, **centroid**, **circumcenter**, and **orthocenter**, to *A*, *B*, *C*. Then add the line determined by *G* and *O*. It should pass through *H*. What is the name of this line?

TO1D. Use **TO1C** to create **Euler line**, which constructs the Euler line of a triangle.

TO1E. Use **SS2I** to create **incenter**, which constructs the incenter and incircle of a triangle.

You should now have at least five tools in a collection that will soon grow considerably.

SECTION 2 # Tools for Special Triangles

Among the triangles derived from a given triangle *ABC*, one of the most often encountered is the one whose vertices are the midpoints of the sides. Write the midpoint of *BC* as *A'*, the midpoint of *CA* as *B'*, and the midpoint of *AB* as *C'*. The triangle *A'B'C'* is known as the *medial triangle* of △*ABC*.

Medial Triangle and Anticomplementary Triangle

We've already encountered the anticomplementary triangle, whose sides are pairwise parallel to the sides of △*ABC*. It is easy to prove that △*ABC* is the medial triangle of its anticomplementary triangle. You can use Sketchpad to confirm this, in the sense of examining hundreds of cases and "seeing" that it "always" works.

But let's digress for a moment to distinguish between "confirm", which is what you can do with Sketchpad, and "prove". To prove means to deduce from postulates. Various sets of postulates (usually rather elaborate) are found in geometry books, along with proofs of propositions implied by postulates. Thus, a proof of a proposition $\mathbb{P}$ is essentially a deductive linkage of already proved or assumed propositions, starting with the postulates and concluding with $\mathbb{P}$. Proof is heavily dependent on language and symbols; pictures are often helpful for understanding a proof, but pictures are not essential parts of a proof. Sketches don't "prove" propositions, but they certainly offer compelling confirmations.

At this point, you should pencil through a proof of your own, using only words and symbols, that △*ABC* is the medial triangle of its anticomplementary triangle. That is, prove that *A* is the midpoint of segment *B''C''*, and similarly for *B* and *C*. A consequence of this theorem is that the sides of the medial triangle are parallel to the matching sides of △*ABC*, as confirmed by Figure 2.2.

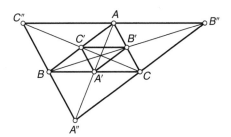

FIGURE 2.2 Triangle *ABC*, its medial triangle and its anticomplementary triangle

ASSIGNMENT 2.1

TO2A. Start with a triangle *ABC*. Construct its medial triangle and save the result as **medial triangle**, including a tool named **medial**.

TO2B. Start with a triangle *ABC*. Construct its anticomplementary triangle and save the result as **anticomplementary triangle.**

Orthic Triangle

The orthic triangle belongs to the orthocenter, *H*, in the same way that the medial triangle belongs to the centroid. Explicitly, let

$$A' = AH \cap BC \quad B' = BH \cap CA \quad C' = CH \cap AB$$

The points *A'*, *B'*, *C'*, often called the *feet of the altitudes* of △*ABC*, are the vertices of the *orthic triangle*.

An interesting connection between the orthic and medial triangles is that they have the same circumcircle. That is, the circle through the vertices of one of these triangles passes through the vertices of the other. This circle could therefore be called the six-point circle. However, its name is the *nine-point circle,* because there are three more notable points on it, namely, the midpoints between *H* and the vertices *A*, *B*, *C*. See Figure 2.3.

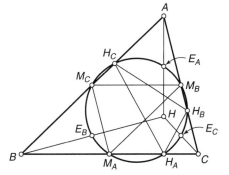

FIGURE 2.3 Triangle *ABC* with its nine-point circle, and the nine points

ASSIGNMENT 2.2

TO2C. Start with a triangle *ABC*. Construct its orthic triangle and save the result as **orthic triangle**, including a tool named **orthic**.

TO2D. Start with a triangle *ABC*. Use your tools to construct both the medial and orthic triangles. Then construct the circle that passes through the vertices of the medial triangle. (Use a tool.) Save the result as **ninepoint circle.**

TO2E. Emulate Figure 2.3, including the same labels for the nine points for which the nine-point circle is named.

TO2F. The points E_A, E_B, E_C in Figure 2.3 are the vertices of the *Euler triangle* of $\triangle ABC$. Figure out how to use **Measure | Length** to confirm that line $E_B E_C$ is parallel to line BC. (The other matching pairs of sidelines are parallel too, so that now we have a nest of four triangles, each pair having pairwise parallel sides.)

TO2G. Use **medial** to confirm by measurements that the four triangles $A'B'C'$ (the medial triangle), $AB'C'$, $A'BC'$, $A'B'C$ are similar to $\triangle ABC$.

TO2H. Use **orthic** to confirm by measurements that the four triangles ABC, AVW, BWU, CUV, where UVW is the orthic triangle of $\triangle ABC$, are similar.

TO2I. Use **orthic** to discover and confirm that the vertex angles of the orthic triangle can be simply expressed in terms of the vertex angles of $\triangle ABC$.

SECTION 3 # Tools for More General Triangles

Having recently worked with the medial and orthic triangles, it has probably occurred to you that you were doing the "same thing" in both cases. In one case, you started with G, and in the other, with H. In both cases, the resulting triangle was inscribed in $\triangle ABC$. Our next venture will be to generalize so as to include both cases and millions of other inscribed triangles of this particular type, called cevian triangles.

Cevian Triangles

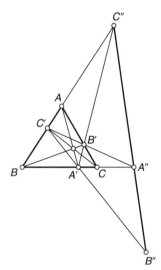

FIGURE 2.4 Desargues's Theorem: if lines AA', BB', CC' concur, then points A'', B'', C'' are collinear

Let P be a point, anywhere in the plane of $\triangle ABC$ except on one of the sidelines, BC, CA, AB. The lines AP, BP, CP, called the *cevians* of P, meet the sidelines in points A', B', C', and the triangle $A'B'C'$ is what we want. It is called the *cevian triangle* of P. (The name honors Giovanni Ceva (1647–1734), an Italian geometer; the surname is pronounced Chay´vah, and the adjective, cheh´vian.)

The line $B'C'$ meets sides AB and AC in an obvious manner, but the point where the line $B'C'$ meets the remaining side, BC, is of particular interest. Call this point A''; if $B'C'$ is parallel to BC then A'' is "at infinity"; in this case, although you can't see A'' on the screen, there is nothing "wrong" with it, and in any case, you can drag P so that A'' becomes visible. Similarly, let $B'' = C'A' \cap CA$ and $C'' = A'B' \cap AB$. According to Desargues's Theorem, A'', B'', C'' are collinear, as indicated by Figure 2.4.

Another highly useful theorem associated with every cevian triangle is Ceva's Theorem. Using the notation of Figure 2.5, this theorem states that

$$|BA'| \cdot |CB'| \cdot |AC'| = |A'C| \cdot |B'A| \cdot |C'B|$$

This is easily remembered by noting that the six lengths go around the perimeter of △*ABC*, the three on the left-hand side alternating with the other three. The converse of Ceva's Theorem is true; i.e., if the distance-product equation holds, then the three lines, *AA'*, *BB'*, *CC'* concur in a point.

Not many kings have published original mathematical proofs. One who did, however, has been identified through the study of recently discovered manuscripts. His name was Yūsuf al-Mu'taman ibn Hūd, King of Saragossa (in Spain) from 1081 to 1085. This king stated and proved Ceva's theorem long before Ceva did in 1678!

TO3A. Create and save **cevian triangle**, for which the givens are points labeled *A*, *B*, *C*, *P* and the result is the *P*-cevian triangle of △*ABC*.

TO3B. Confirm Desargues's Theorem using labels that match the above description. (It's up to you to find a way to use **Measure | Calculate** to confirm that *A''*, *B''*, *C''* are collinear.)

TO3C. Emulate Figure 2.5, including the measurements.

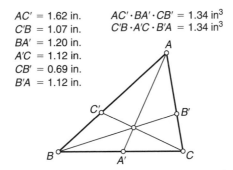

$AC' = 1.62$ in. $AC' \cdot BA' \cdot CB' = 1.34$ in^3
$C'B = 1.07$ in. $C'B \cdot A'C \cdot B'A = 1.34$ in^3
$BA' = 1.20$ in.
$A'C = 1.12$ in.
$CB' = 0.69$ in.
$B'A = 1.12$ in.

FIGURE 2.5 Ceva's Theorem

TO3D. Start with △*ABC*. Sketch a movable point *B'* on side *AC* and a movable point *C'* on side *AB*. Let *P = BB' ∩ CC'*. Sketch a movable point *A'* on side *BC*. Use **Measure | Calculate** to print the products,

$$|BA'| \cdot |CB'| \cdot |AC'| \quad \text{and} \quad |A'C| \cdot |B'A| \cdot |C'B|$$

Confirm that these two are equal only when line *AA'* passes through *P*. In your caption, tell how this result relates to Ceva's Theorem.

Harmonic Conjugates

Recall the anticomplementary triangle, as in Figure 2.2. We've already seen that it's an "opposite" of the medial triangle. Now we seek an "opposite" for an *arbitrary* cevian triangle. That is, given an independent point *P* inside (for temporary convenience) △*ABC*, we can sketch the cevian triangle of *P*, which is inscribed in △*ABC*. The new triangle we're looking for must circumscribe △*ABC* and have vertices on the lines *AP*, *BP*, *CP*. You should try this out with pencil and paper. For some choices of △*ABC* and *P*, you may find it surprising that the desired triangle exists. If you discover how to construct it on your own, that's quite good!

This quest leads to the notion of an *anticevian triangle* of a point. In order to construct anticevian triangles, we must first become acquainted with harmonic conjugates—which will have many other applications as well.

Suppose U and V are points, and P is a point on line UV. The *harmonic conjugate of P with respect to U and V* is the point Q on line UV that satisfies

$$\frac{|UQ|}{|VQ|} = \frac{|UP|}{|VP|}$$

Harmonic conjugates come in pairs—each point (P and Q) is the harmonic conjugate of the other.

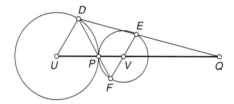

FIGURE 2.6 Q is the harmonic conjugate of P with respect to U and V

A construction of Q from given collinear points U, V, P is indicated by Figure 2.6. Let D be a point on $\circ(U, |UP|)$, let L be the line through V parallel to UD, let E be the point as shown where L meets $\circ(V, |VP|)$; then $Q = DE \cap UV$.

To prove that Q is the harmonic conjugate of P with respect to U and V, let F be the point, other than E, where L meets $\circ(V, |VP|)$. Then $\triangle QUD$ is similar to $\triangle QVE$, and $\triangle DUP$ is similar to $\triangle PFV$. Consequently,

$$\frac{|UQ|}{|VQ|} = \frac{|UD|}{|VE|} \quad \text{and} \quad \frac{|UP|}{|VP|} = \frac{|UD|}{|VF|}$$

Now, since segments VF and VE are radii of a circle, the proof is finished.

ASSIGNMENT 3.2

TO3E. Sketch two points, U and V, and their line. If you place a movable point P on line UV, between U and V, you'll find that Sketchpad won't accept P as a given when you try to create a tool. This is because P is not an independent point. To get around this obstacle, we must construct P from U, V and a point P' that determines P. Perhaps the simplest way is to take as P' a point on the line perpendicular to line UV. Sketchpad will then accept P' as a given (unless $P' = P$), and you can construct P from P' and continue the construction given above, using for D a point where line UP meets the circle centered at U. Save your sketch, including a tool named **harmonic conjugate**.

TO3F. Use **harmonic conjugate** using **Calculate | Measure** to confirm the fact that if P and Q are harmonic conjugates with respect to U and V, then U and V are harmonic conjugates with respect to P and Q.

Anticevian Triangles

Suppose ABC is a triangle and P is a point not on one of the sidelines BC, CA, AB. Let $A'B'C'$ be the cevian triangle of P. Let A'' be the harmonic conjugate of P with respect to A and A'. Let B'' be obtained likewise from B, B', P, and C'' from C, C', P. The *anticevian triangle* of P is the triangle $A''B''C''$.

Several triangles that have been studied for centuries are good examples of anticevian triangles. Already mentioned is the one for which P is the centroid, G. If $P = I$, the resulting anticevian triangle is the *excentral triangle,* since the vertices are the excenters of $\triangle ABC$. A third example is the *tangential triangle,* which is the anticevian triangle of the symmedian point. As this point has not been introduced here previously, we'll construct it next: Reflect the median AG about the bisector of $\angle CAB$, reflect BG about the bisector of $\angle ABC$, and reflect CG about the bisector of $\angle BCA$. The three reflected lines concur in the *symmedian point, K,* as shown in Figure 2.7.

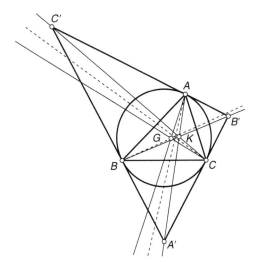

FIGURE 2.7 Triangle ABC with its symmedian point, K, and tangential triangle, $A'B'C'$

ASSIGNMENT 3.3

TO3G. Starting with $\triangle ABC$ and a point P, construct the anticevian triangle of P. Save this as **TO3G**, including the tool **anticevian triangle**. (Construct A'', B'', and C'' directly as in Figure 2.6; do not try to apply your **harmonic conjugate** tool.)

TO3H. Apply **anticevian triangle** to A, B, C, I to obtain the anticevian triangle, $A'B'C'$, of the incenter. Be sure A' is the label of the vertex that matches A. Construct the segment from A' perpendicular to line BC, and then construct the circle having this segment as radius and A' as center. Is this circle tangent to all three sides of $\triangle ABC$? How many circles are tangent to all three sides?

TO3I. Follow the steps presented above to construct the symmedian point, K, of $\triangle ABC$. Use **anticevian triangle** to construct the anticevian triangle, $A'B'C'$, of K. Construct the circumcircle of $\triangle ABC$ and segments KA, KB, KC. Does $A'B'C'$ appear to deserve its name, the tangential triangle? (Save your construction of K as **symmedian point.**)

TO3J. Use **cevian triangle** and **anticevian triangle** to construct the orthic and tangential triangles of $\triangle ABC$. Find a way to use **Measure | Calculate** to confirm that the sides of these triangles are pairwise parallel.

TO3K. Starting with $\triangle ABC$ and an independent point P, construct the cevian triangle $A'B'C'$ of P, and then the anticevian triangle $A''B''C''$ of P. Use **Measure | Calculate** to check that the distance ratios that define harmonic conjugates hold for the points A, P, A', A''.

Perspective Triangles

Suppose you have two triangles and that it is possible to label the vertices of one as R, S, T and the other as U, V, W in such a way that the lines RU, SV, TW concur in a point, P. Then the triangles are called *perspective triangles,* and each is perspective to the other. The point P is the *perspector.* By Desargues's Theorem, the points

$$ST \cap VW \quad TR \cap WU \quad RS \cap UV$$

are collinear; their line is the *perspectrix.* (In older literature *perspector* is "center of perspective" and *perspectrix* is "axis of perspective". Also, triangles were "in perspective". The modern usage, typified by the phrase "$\triangle RST$ is perspective to $\triangle UVW$" conforms to other usages, such as "is similar to" and "is congruent to".)

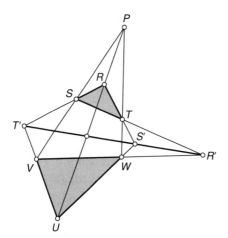

FIGURE 2.8 Perspective triangles, along with their perspector and perspectrix

Figure 2.8 shows perspective triangles RST and UVW along with their perspector and perspectrix. Some earlier figures also provide examples: every cevian triangle of $\triangle ABC$ is perspective to $\triangle ABC$; hence, every anticevian triangle is also perspective to $\triangle ABC$. Soon we'll see that much more is true: *every cevian triangle is perspective to every anticevian triangle!*

ASSIGNMENT 3.4

TO3L. Starting with $\triangle ABC$ and independent points P and Q, construct the cevian triangle $A'B'C'$ of P and the anticevian triangle $A''B''C''$ of Q. Confirm the assertion that $\triangle A'B'C'$ is perspective to $\triangle A''B''C''$. The perspector is called the *P-Ceva conjugate of* Q. Create and save a tool, **Ceva conjugate**, that constructs from givens A, B, C, P, Q the P-Ceva conjugate of Q.

TO3M. Loosely speaking, the word *conjugate* is used for a function f satisfying $f(f(x)) = x$. Confirm, by applying **Ceva conjugate** twice, that Ceva-conjugacy really is a conjugacy. In other words, confirm that for any P and Q, the P-Ceva conjugate of the P-Ceva conjugate of Q is Q.

Pedal Triangles

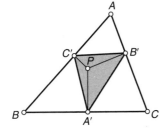

FIGURE 2.9 Pedal triangle of a point

Suppose ABC is a triangle and P is a point not lying on a sideline of $\triangle ABC$. Let A' be the point in which the line through P perpendicular to line BC meets line BC. Let B' be the analogous point on line CA, and C' that on line AB, as shown in Figure 2.9. The triangle $A'B'C'$ is the *pedal triangle* of P.

Every time that the familiar phrase "the lines AA', BB', CC' concur" applies, we have an example of perspective triangles. Perhaps you suspect that every "reasonable" triangle $A'B'C'$ is perspective to $\triangle ABC$. We'll soon see, however, that this is not the case for a pedal triangle of the centroid.

TO3N. Starting with $\triangle ABC$ and its circumcircle Γ, let P be an independent point, and sketch the lines PA, PB, PC. Label as A' the point other than A in which line PA meets Γ, and likewise, construct and label points B' and C' on Γ. Use **Measure** to confirm that $\triangle A'B'C'$ is similar to the pedal triangle of P by comparing corresponding angles. Then use **Measure | Calculate** to further confirm the similarity by comparing ratios of corresponding sidelengths.

TO3O. Use Sketchpad to demonstrate that for many triangles $\triangle ABC$, the pedal triangle of the centroid is not perspective to $\triangle ABC$.

Antipedal Triangles

You can make a good guess about the meaning of "antipedal triangle" by recalling the connection between cevian and anticevian triangles. The *antipedal triangle* of P is the triangle whose P-pedal triangle is the reference triangle, $\triangle ABC$. To construct this new triangle, start with the line through A perpendicular to line AP. Similarly, sketch the line through B perpendicular to BP and the line through C perpendicular to CP. These last two lines intersect in a point A'', and the other two likewise determine points B'' and C''. Triangle $A''B''C''$ is the antipedal of P. A quick penciled sketch will show that the P-pedal triangle of $A''B''C''$ is indeed $\triangle ABC$. Figure 2.10 shows the antipedal of the centroid, G, together with the pedal triangle of the symmedian point, K.

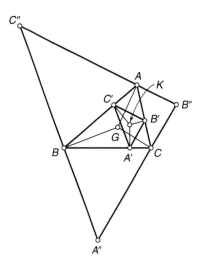

FIGURE 2.10 Antipedal triangle $A''B''C''$ of G and pedal triangle $A'B'C'$ of K

ASSIGNMENT 3.6

TO3P. Use **Measure | Calculate** to confirm that the antipedal triangle of K is similar to the pedal triangle of G.

TO3Q. Use **Calculate** to confirm that the antipedal triangle of G is similar to the pedal triangle of K.

Isogonal Conjugates

To construct the isogonal conjugate of a point P not on a sideline BC, CA, AB of $\triangle ABC$, reflect the line AP about the bisector of $\angle CAB$, reflect BP about the bisector of $\angle ABC$, and reflect CP about the bisector of $\angle BCA$. The three reflected lines concur in the desired isogonal conjugate of P, shown as point Q in Figure 2.11.

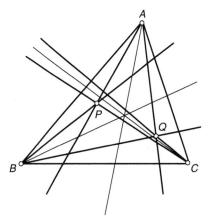

FIGURE 2.11 Reflected lines AP, BP, CP concurring in Q, the isogonal conjugate of P

Isogonal conjugacy is the most common kind of conjugacy in triangle geometry. Later we'll need a tool that constructs the isogonal conjugate of a point.

TO3R. Starting with $\triangle ABC$ and a point P, construct the isogonal conjugate, labeled P', of P. Create **isogonal conjugate** as a tool.

TO3S. Continuing from **TO3R**, measure the distances from P to the sidelines, BC, CA, AB, and also measure the distances from P' to the sidelines. If UVW is the pedal triangle of P and $U'V'W'$ is the pedal triangle of P', then your six measurements can be written as

$$|PU| \quad |PV| \quad |PW| \qquad |P'U'| \quad |P'V'| \quad |P'W'|$$

Use **Measure | Calculate** to confirm that

$$|PU| \cdot |P'U'| = |PV| \cdot |P'V'| = |PW| \cdot |P'W'|$$

That is, the distances from P are inversely proportional to the corresponding distances from P'. Next, drag P to a location for which the six distances are equal. Finally, drag P to a location for which the first three distances are respectively proportional to the sidelengths $|BC|$, $|CA|$, $|AB|$. As always, record your findings in a caption.

TO3T. Starting with $\triangle ABC$ and an arbitrary point P, let Δ, Δ', Δ'' denote the areas of $\triangle ABC$, the pedal triangle of P, and the antipedal triangle of the isogonal conjugate of P, respectively. Use **Measure | Calculate** to confirm that

$$\Delta' \cdot \Delta'' = \Delta^2$$

That is, calculate and print the product on the left-hand side, calculate and print Δ^2, and observe that as you drag A, B, C, P, the two calculations remain equal.

TO3U. Continuing **TO3T**, delete the measurements, and then use **Measure | Calculate** to confirm that the sides of the pedal triangle of P are pairwise parallel to those of the antipedal triangle of the isogonal conjugate of P.

SECTION 4 # Sliders and Lines

Frequently it is desirable to control one or more numerical variables, such as the slope of a line. Such control is easily provided by sliders. In this section, we'll start with a single slider for a variable m, and use it to vary the slope of the line $y = mx$.

The length of a segment with a movable endpoint can serve as a slider. However, the values provided by such a slider cannot be negative, since length cannot be negative. We certainly want to enable the slope m to be negative. So, the key idea for a slider is that its values will be the x-coordinate (abscissa) or y-coordinate (ordinate) of a movable endpoint, relative to the other endpoint, which will have coordinate 0.

In order to graph the line $y = mx$, we'll need an xy-coordinate system. The slider for m could use the origin of that coordinate system, but it works better to place a second origin away from the first.

1. Start with **Graph | Define Coordinate System** and **Graph Grid Form | Square**. Place a point somewhere away from the origin, select it, and apply **Graph | Define Origin**. Label the first origin with the letter **O** and the other origin with the number **0**. Notice that both *x*-axes have a unit point labeled **1**. You can drag these unit points to adjust the corresponding scales.

2. To keep the screen fairly clear, apply **Graph | Hide Grid**. When you first apply this command, the "active" grid will be cleared; to clear the other grid, select the other origin, apply **Graph | Mark Coordinate System**, and **Graph | Hide Grid**. You can also hide axes in the usual way. For now, hide the *y*-axis of the remote system.

3. Place an independent point on the remote *x*-axis and label it **m**. Construct the segment from **0** to **m**, and then hide the *x*-axis containing segment **0m**. Before borrowing the *x*-coordinate of **m**, you must be sure that this coordinate will be measured in the correct coordinate system, so select the remote **0** and mark the coordinate system. Then select **m** and apply **Measure | Abscissa**. Drag **m** back and forth across **0** and check that the printed measure takes both positive and negative values. Note that the unit point is still visible, even though the *x*-axis is hidden. Label it with a **1**.

4. It is customary to use boldface print for the labels of **0, 1**, and **m** and to apply **Display | Line Width | Thick** to segment **0m**. Another refinement is to construct thin-width segments between **0** and **1** and between **m** and **1**, so that the collinearity of **0, 1**, and **m** will remain always apparent. The printed value of the abscissa of **m** appears as x_m.

You now have a slider for a variable, in this case, the slope *m*. The steps above can be used to create several sliders at the beginning of many very useful sketches.

In order to use the value of *m* given by the slider, apply **Graph | New Function**. A dialog box appears. Select x_m and note that it then appears in the dialog box. Either type an asterisk, or click the asterisk in the dialog box. Then type **x** or click the **x** in the dialog box. Click **OK**, and note that an equation for your function is now selected on the screen. Once again, change coordinate systems; that is, select the original origin and apply **Graph | Mark Coordinate System**. Then apply **Graph | Plot New Function**. Now for the final payoff: drag **m** and see that the line $y = mx$ rotates about the origin. Check that when $m = 0$, your line is horizontal.

ASSIGNMENT 4.1

TO4A. Create **line y = mx** by following the steps given above. Confirm as in Figure 2.12 that the slope y/x equals the tangent of the angle between the positive *x*-axis and the line $y = mx$. Add a movable point (x, y) and its children *x* and *y* as in Figure 2.12. (The tangent function is built-in on Sketchpad's **Calculator**.)

TO4B. Augment the sketch for **TO4A** by introducing a second slider for *y*-intercept, denoted by *b*, and create and save a sketch named **line y = mx + b**.

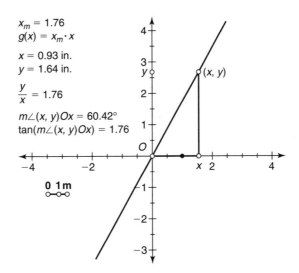

$x_m = 1.76$
$g(x) = x_m \cdot x$

$x = 0.93$ in.
$y = 1.64$ in.

$\dfrac{y}{x} = 1.76$

$m\angle (x, y)Ox = 60.42°$
$\tan(m\angle (x, y)Ox) = 1.76$

FIGURE 2.12 The line $y = mx$. Drag m to vary the slope.

TO4C. Augment the sketch for **TO4B** by introducing a third slider and relabeling so as to graph the line $ax + by + c = 0$, where a, b, c are controlled by sliders. Save your sketch as **line ax + by + c = 0**.

PROJECTS

PROJECT 1: DOUBLE PERSPECTIVE

Recall from Section 3 that triangles *ABC* and *UVW* are perspective if there is a way to match vertices *A*, *B*, *C* with vertices *U*, *V*, *W* using concurrent lines. There are six possible matches, as indicated by

AU, BV, CW	*AU, BW, CV*
AV, BW, CU	*AV, BU, CW*
AW, BU, CV	*AW, BV, CU*

and the triangles are *doubly perspective* if two of these triples of lines concur.

Part 1. Use a pencil and paper to devise an example of triangles *ABC* and *UVW* that are doubly perspective. Then construct such triangles using Sketchpad. Include three red lines concurring in a perspector and three blue lines concurring in another perspector.

Part 2. Continuing from **Part 1**, construct the perspectrix (cf. Figure 2.8) corresponding to one of the perspectors.

PROJECT 2: TRIPLE PERSPECTIVE

This project begins with the construction represented by Figure 2.13.

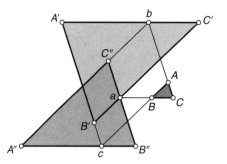

FIGURE 2.13 Triple perspective triangles $A'B'C'$ and $A''B''C''$

Part 1. Start with a triangle ABC, and construct two triangles as follows.

1. Use **circumcenter** to sketch the circumcenter, O, of $\triangle ABC$

2. Let $a' = AO \cap BC$

3. Use **harmonic conjugate** to sketch the harmonic conjugate, a, of a' with respect to B and C

4. Similarly, sketch b and b' on line CA and c and c' on line AB

5. Sketch lines through a, b, c parallel to sidelines AB, BC, CA, respectively, and label the resulting triangle $A'B'C'$ as in Figure 2.13

6. Sketch lines through a, b, c parallel to sidelines AC, BA, CB, respectively, and label the resulting triangle $A''B''C''$ as in Figure 2.13

Part 2. Confirm using Sketchpad that three of the six possible matchings of vertices of $A'B'C'$ to vertices of $A''B''C''$ are perspective matchings.

Part 3. In how many ways is $\triangle A'B'C'$ perspective to $\triangle ABC$?

Part 4. According to Desargues's theorem, for each perspector there is also a perspectrix. Sketch the perspector and perspectrix for one of the three perspectives in **Part 2**.

Part 5. Sketch a perspector and perspectrix for the perspective triangles in **Part 3**.

PROJECT 3: CENTROIDS

Sketch arbitrary triangles ABC and UVW, and apply **centroid** to sketch their centroids. The centroid of the two triangles, taken together, is a balance-point, or center of mass, for the set of all points in both triangles. Let G_1, G_2, and G denote the centroids of $\triangle ABC$, $\triangle UVW$, and the two triangles together, respectively. Then G lies on segment G_1G_2. In fact, its location is easily formulated: let

$$\mathcal{A}_1 = \text{area}(ABC) \quad \text{and} \quad \mathcal{A}_2 = \text{area}(UVW)$$

then G is the point on segment G_1G_2 that satisfies

$$|G_1G|/|G_1G_2| = \mathcal{A}_2/(\mathcal{A}_1 + \mathcal{A}_2)$$

Part 1. Sketch arbitrary triangles ABC and UVW and their respective centroids, G_1 and G_2. Let T_1 denote the region consisting of all the points inside and on $\triangle ABC$, and let T_2 denote the region consisting of all the points inside and on $\triangle UVW$, so that $T_1 \cup T_2$ denotes the region consisting of both triangles and their interiors. Figure out how to construct the centroid, G, of the region $T_1 \cup T_2$. (You might wonder if it is necessary to include the interiors of the triangles. The answer is yes. The centroid of the triangle itself—meaning just the sides—is a special point called the *Spieker center* of the triangle.) Drag points in your sketch so that it resembles Figure 2.14, and save it.

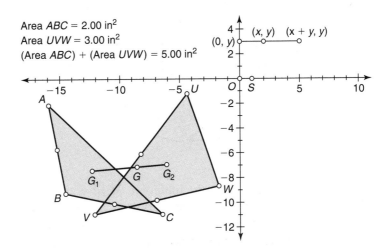

Area ABC = 2.00 in^2
Area UVW = 3.00 in^2
(Area ABC) + (Area UVW) = 5.00 in^2

FIGURE 2.14 Centroid, G, of two triangles

Part 2. Where do you think G will be when $G_1 = G_2$? Confirm your answer by dragging the triangles appropriately. (To drag a whole triangle, simply drag its centroid.) Where do you think G will be when $\mathcal{A}_1 = \mathcal{A}_2$? Confirm that G is quite close to G_2 when $\mathcal{A}_1$ is much smaller than $\mathcal{A}_2$.

Part 3. Sketch the centroids G_1, G_2, G_3 of three arbitrary triangles T_1, T_2, T_3 (including interiors), respectively. Then sketch the centroid $G_{1,2}$ of $T_1 \cup T_2$, the centroid $G_{1,3}$ of $T_1 \cup T_3$, and the centroid $G_{2,3}$ of $T_2 \cup T_3$. Sketch the centroid G of $T_1 \cup T_2 \cup T_3$. Must G lie inside $\triangle G_1G_2G_3$?

Part 4. Create a sketch that shows the centroids of three arbitrary circular disks, the centroids of the unions of the disks taken two at a time, and the centroid of the union of all three disks. Use dilations, not coordinates.

PROJECT 4: DESARGUES'S THEOREM AND ITS CONVERSE

Figure 2.4 illustrates Desargues's theorem in a special case; that point A' can be any point on line BC, not necessarily between B and C. The converse of the theorem is true: if A'', B'', C'' are collinear, then AA', BB', CC' concur. Any figure illustrating Desargues's theorem also illustrates its converse. In any such figure, the perspector is the point of concurrence of the lines AA', BB', CC', and the perspectrix is the line of the points A'', B'', C''. This project will explore the theorem and its converse when the perspector is a movable point on a circle or a line.

Part 1. Start with a triangle ABC and an arbitrary circle Γ. Let P be a movable point on Γ. Sketch

$$A' = AP \cap BC \quad B' = BP \cap CA \quad C' = CP \cap AB$$

Sketch A'', B'', C'' as in Figure 2.15. Then sketch line $B''C''$.

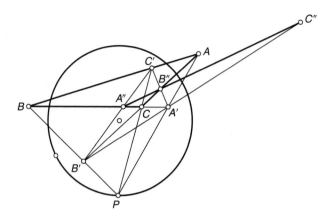

FIGURE 2.15 The configuration of Figure 2.4, with perspector P movable on a circle

Part 2. Sketch the midpoint, M, of segment $B''C''$. Select P and M, and apply **Construct | Locus**. Vary P, Γ, and $\triangle ABC$. As you can see, the locus of M is not a simple curve. (If your locus is choppy, apply **Edit | Select All** followed by **Edit | Advanced Preferences** so that you can type a larger number into the appropriate box, and then re-apply **Locus**)

Part 3. Repeat **Part 1**, except this time provide an arbitrary line for P to move on, instead of Γ.

Part 4. Start a new sketch with an arbitrary triangle ABC, and an independent line L. Let A'', B'', C'' be the points where L meets lines BC, CA, AB, respectively. Construct points A', B', C' so as to illustrate the converse of Desargues's theorem. Construct the perspector of triangles ABC and $A'B'C'$.

CHAPTER **THREE**

Locus

A *LOCUS* IS a path of a point that moves in accord with certain conditions. Examples of simple loci which you have probably seen before and which you should be able to identify quickly are these:

- the locus of a point whose distance from a fixed point stays constant
- the locus of a point that keeps equal distances from two given lines

  Less common loci are these:

- the locus of a tack on a wheel rolling on a line
- the locus of a point Y for which $|YA| = 2|YB|$, where points A and B stay fixed
- the locus of a point Y for which $|YF| = |YD|$, where the point F and line D stay fixed

  This last example is a parabola, one of the members of a family called conic sections.

SECTION 1 Conic Sections

Among the approaches to the subject of conic sections are two very common ones that we'll discuss in this section. One approach accounts for the name, which refers to a cone in 3-dimensional space, and the other refers to a fixed point and a line in 2-dimensional space.

We'll describe the first approach only briefly: imagine (that is, make an *image* in your mind) a cone being sliced by a plane. The points lying on both the cone and plane comprise a curve, and the shape of the curve depends on the angle between the cone's axis and the plane: if the angle is 90°, the curve is a circle; if the plane is then tilted a bit, the circle becomes an ellipse, and as you tilt the plane a bit more, the ellipse becomes more elongated.

If tilted even further, the plane becomes parallel to the "edge" of the cone; the ellipse suddenly snaps open and is no longer an ellipse but a parabola. As you tilt the plane still further, the curve of intersection with the cone is a branch of a hyperbola. To summarize, the three kinds of curve, *ellipse, parabola,* and *hyperbola* have been obtained as sections of a cone, hence the name, *conic section.* We'll turn now to conic sections as defined in a plane.

Parabola

Suppose F is a point and D is a line that does not pass through F. The letters F and D match the names *focus* and *directrix.* We seek the locus of a point whose distances from F and D are equal. Certainly one such point is the one halfway between F and D. Suppose L is a line perpendicular to D, as in Figure 3.1.

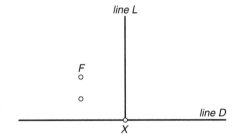

FIGURE 3.1 Ready to find a point on line L that is equidistant from F and D

Imagine a movable point Y on L that starts at the point X where lines L and D meet. When $Y = X$, we have $|YD| = |YX| = 0$, so that $|YD|$ is certainly less than $|YF|$. However, as Y moves on L away from X, the distance $|YD|$, alias $|YX|$, increases, so that Y will eventually reach a position where $|YD| = |YF|$. That would mean that $\triangle XYF$ is isosceles, as in Figure 3.2. Then, as X varies, we'll have a Y for each X. Several such points Y are shown in Figure 3.3.

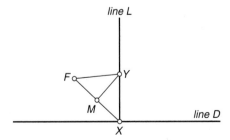

FIGURE 3.2 A point Y equidistant from point F and line D

The construction in Figure 3.3 is ideal for understanding **Locus**. Each position of X yields an image Y. In other words, Y is a function of X; you can write "$Y = f(X)$", where f is the "rule" defined by the method of construction. Another way to say this is that X is the *drive-point* and Y is the *image*. Accordingly, select X and Y, in that order; then apply **Construct | Locus**.

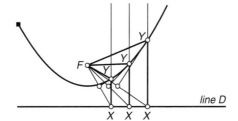

FIGURE 3.3 Locus of point Y equidistant from point F and line

If, when you use **Locus**, the result looks like a polygon, you can change it by increasing the number of points that Sketchpad uses to sample the locus; this is accomplished with **Edit | Advanced Preferences | Sampling**, accessed by clicking **Edit**, pressing **Shift**, and clicking **Advanced Preferences** while keeping **Shift** down. If necessary, delete the previous locus and apply **Construct | Locus** again.

ASSIGNMENT 1.1

LO1A. Construct a parabola using **Locus** and create a tool named **parabola**. Print measures of $|YD|$ and $|YF|$. Use **Measure | Calculate** to print the ratio $|YD|/|YF|$. As always, print your observations in a caption.

LO1B. After saving **LO1A** and **parabola**, predict the loci of the midpoints of segments YF, YX, and XF, and then add these to **LO1A**.

LO1C. In **LO1A**, let F' be the reflection of F in D. Construct the parabola having focus F' and directrix D. Is it possible for the union of two such parabolas to be a hyperbola?

Ellipse

Suppose F_1 and F_2 are points. We seek the locus of a point Y that moves in such a way that the sum of distances $|YF_1| + |YF_2|$ stays constant. Quick work with a pencil should convince you that the least constant possible would be for Y lying on the segment $F_1 F_2$, and this constant would be $|F_1 F_2|$; in this case, however, the locus is simply the segment $F_1 F_2$, so we stipulate that the constant can be any number greater than $|F_1 F_2|$. Write the constant as $2a$. Let C be the midpoint of segment $F_1 F_2$; let V_1 be the point a units from C in the direction of F_1, and let V_2 be the point a units from C in the direction of F_2, so that $|V_1 V_2| = 2a$, as in Figure 3.4.

FIGURE 3.4 Ready to sketch an ellipse

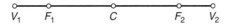

If we regard Y as a variable in the equation $|YF_1| + |YF_2| = 2a$, then two solutions are $Y = V_1$ and $Y = V_2$. The locus we seek therefore includes

V_1 and V_2. In order to find all other solutions, let X be a movable point on segment $V_1 V_2$. Let O_1 be the circle with center F_1 passing through X, and let O_2 be the circle with center F_2 and radius $2a - |XF_1|$. As long as X moves so that $|XF_1| < 2a$, any point Y on O_1 has distance $|YF_1|$ from F_1, and any point Y on O_2 has distance $|YF_2|$ from F_2. Since

$$|YF_1| = |XF_1| \quad \text{and} \quad |YF_2| = 2a - |XF_1|$$

we have, for any Y on *both* circles, the desired sum $|YF_1| + |YF_2| = 2a$. Typical circles O_1 and O_2 are shown in Figure 3.5. The points F_1 and F_2 are the *foci,* the points V_1 and V_2 are the *vertices,* and segment $V_1 V_2$ is the *major axis.*

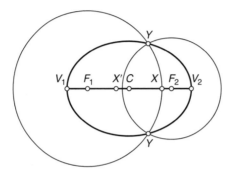

FIGURE 3.5 Locus of point Y satisfying $|YF_1| + |YF_2| = 2a$

To finish the discussion, we must contruct the circle O_2. In Figure 3.5, the point X' is placed so that segment $V_1 X'$ has length $2a - |XF_1|$. In order to construct X', note that

$$|V_1 F_1| + |F_1 X| + |XV_2| = 2a$$

so that

$$2a - |XF_1| = |V_1 F_1| + |XV_2|$$

Thus, X' is the result of translating point F_1 by vector XV_2. You can now select point F_2 and segment $V_1 X'$, and apply **Construct | Circle By Center+Radius**.

ASSIGNMENT 1.2

LO1D. Construct an ellipse using **Locus** (cf. Figure 3.5). Use **Measure | Calculate** to print the sums $|V_1 F_1| + |V_1 F_2|$ and $|YF_1| + |YF_2|$. The latter, as Y goes around the ellipse, should stay equal to the former. Save **LO1D** and a tool named **ellipse**, for which the givens are the vertices. (An independent point F_1 will appear on its own when you apply **ellipse**; after F_1 has appeared, you can drag it.)

LO1E. Continuing from **LO1D**, construct the reflection F_1' of F_1 in V_1, the reflection F_2' of F_2 in V_2, and apply **ellipse** to sketch the ellipse that has foci F_1 and F_2 and passes through F_1' and F_2'. (Sketchpad enables reflection in a point with **Transform | Rotate.**)

LO1F. Starting with **LO1D**, sketch an ellipse whose major axis passes through the midpoint of segment $F_1 F_2$ and is perpendicular to the line $F_1 F_2$.

LO1G. Sketch a circle $\circ O$ and two points P and Q outside $\circ O$. Imagine that Daniel needs to meet Josepha on $\circ O$ and then continue to Q in the least possible time. Figure out how to use ellipses with foci P and Q to determine where Daniel and Josepha should meet. (You needn't actually construct the point; just explain in a caption how to use your sketch to locate it.)

Hyperbola

Suppose F_1 and F_2 are points. The locus of a point Y that moves in such a way that the $|YF_1| - |YF_2|$ stays constant, $2a$, is one branch of a hyperbola; the other branch is the locus of Y satisfying $|YF_2| - |YF_1| = 2a$. Thus, the complete hyperbola is the locus of Y satisfying $|YF_2| - |YF_1| = \pm 2a$, or, equivalently, $||YF_2| - |YF_1|| = 2a$. See Figure 3.6.

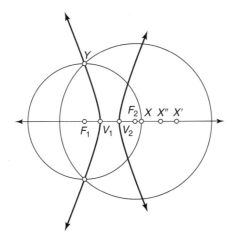

FIGURE 3.6 A hyperbola, the locus of point Y satisfying $|YF_2| - |YF_1| = \pm 2a$

ASSIGNMENT 1.3

LO1H. Figure out a construction for a hyperbola, based on the preceding construction of an ellipse. Save **LO1H** and a tool named **hyperbola**, for which the givens are two vertices and one focus, or else two foci and one vertex.

LO1I. Start with **LO1D**. Add the hyperbola whose vertices are F_1 and F_2 and whose foci are V_1 and V_2.

Eccentricity

Suppose D is a line and F is a point not on D. The locus of a point Y such that $|YF| = |YD|$ is a parabola. Written in the form $|YF|/|YD| = 1$, this equation leads to an interesting question: what is the locus of P satisfying $|YF|/|YD| = e$, where e is a positive number that isn't necessarily 1? The answer is that the locus of Y is an ellipse if $e < 1$, a parabola if $e = 1$, and a hyperbola if $e > 1$. If $e = 0$, the ellipse is a circle. In all cases the number e is called the *eccentricity* of the conic. (From here on, we'll write *conic* for *conic section*.)

We'll now set out to construct the conic determined by any given D, F, and e. The main new thing here is what to do with e. We want it to be a predetermined ratio, $|YF|/|YD|$. One way to handle this is to carry out an auxiliary construction in the corner of the screen. For example, sketch a horizontal segment AB, and place a point U on it. Construct a segment BC perpendicular to AB, and sketch $\triangle ABC$. Then construct a line through U parallel to BC. Let V be the point where this line meets segment AC. You now have similar triangles AUV and ABC, so that the ratios of segment lengths, $|UV|/|BC|$ and $|AU|/|AB|$, are equal. Given any e between 0 and 1, you can drag U so that

$$|AU|/|AB| = e$$

Select segments UV and BC, in that order, and apply **Transform | Mark Segment Ratio**.

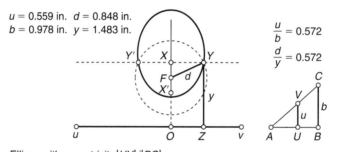

$u = 0.559$ in. $d = 0.848$ in.
$b = 0.978$ in. $y = 1.483$ in.

$\dfrac{u}{b} = 0.572$

$\dfrac{d}{y} = 0.572$

Ellipse with eccentricity $|UV|/|BC|$.
F = focus, line uv = directrix; point Y keeps $|FY|/|ZY|$ = eccentricity.
Move X on line OF to drive Y around ellipse.
Move U on line AB to vary eccentricity of ellipse.

FIGURE 3.7 Ellipse with user-controlled eccentricity

Figure 3.7 indicates how to use the marked ratio to sketch an ellipse. Start with a line D and a point F not on D. Point F will be the focus and D the directrix. Line OF is perpendicular to D, and X is a movable point on line OF. Construct a dashed line D' through X parallel to line D. Select point O and apply **Transform | Mark Center**. Then select point X, and apply **Transform | Dilate**. The result is a point X' satisfying

$$|OX'|/|OX| = e$$

Select point F and segment OX', and apply **Construct | Circle By Center+Radius**.

Let Y and Y' be the points of intersection of the circle and line D'. Being on the circle, point Y has distance $|OX'|$ from F, and being on line D', point Y has distance $|OX|$ from line D. Since $|OX'|/|OX| = e$, we have

$$|YF|/|YD| = e$$

as desired.

ASSIGNMENT 1.4

LO1J. Emulate Figure 3.7. Observe that as you vary U near A, the ratio e is small and your ellipse is nearly circular. Then as you vary U closer to B, the eccentricity e approaches 1, and your ellipse becomes longer. When $U = B$, you have $e = 1$, and the locus has snapped to a different kind of curve.

L01K. Starting with **L01J**, delete X'. Select, in your "ratio-triangle", segments BC and UV, in that order, so that the ratio e exceeds 1. Follow the same construction as before, this time obtaining a hyperbola. Thus, **L01J** and **L01K** enable you to confirm that the conic is an ellipse, parabola, or hyperbola according as e is less than 1, equal to 1, or greater than 1.

SECTION 2 Conics in Analytic Geometry

In the xy-plane, the conics have standard positions and matching standard forms of equations. For ellipses, the standard form is

$$\frac{x^2}{a^2} + \frac{y^2}{b^2} = 1$$

and for hyperbolas the standard forms are

$$\frac{x^2}{a^2} - \frac{y^2}{b^2} = 1 \quad \text{and} \quad \frac{y^2}{a^2} - \frac{x^2}{b^2} = 1$$

We'll use Sketchpad to plot these curves. Then we'll plot parabolas of the form $y = ax^2 + bx + c$; that is, parabolas that open vertically.

Ellipse in Standard Form: $\frac{x^2}{a^2} + \frac{y^2}{b^2} = 1$

Obviously, the four points $(a, 0)$, $(-a, 0)$, $(0, b)$, $(0 - b)$ are on this ellipse. If $a > b$, the major axis of the ellipse lies within the x axis. Letting $c = \sqrt{a^2 - b^2}$, the foci are the points $(-c, 0)$ and $(c, 0)$. Thus, taking

$$V_1 = (-a, 0) \quad V_2 = (a, 0) \quad F_1 = (-c, 0) \quad F_2 = (c, 0)$$

we can apply **ellipse** (see **L01D**) to sketch this ellipse. The eccentricity of this ellipse is c/a. See Figure 3.8.

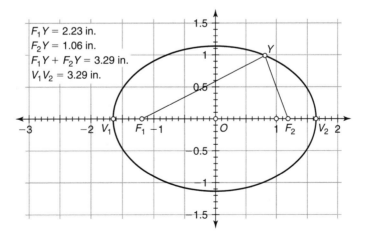

FIGURE 3.8 Ellipse in standard position

LO2A. Start with **File | New Sketch**, and display a square grid. Place points V_1 and V_2 on the x axis so the the origin is their midpoint. Apply **ellipse** and then adjust your sketch so that it resembles Figure 3.8.

LO2B. If $b > a$, the major axis of the ellipse is vertical. Adapt the procedure in **LO2A** to obtain an ellipse with vertical major axis. The equation $c = \sqrt{b^2 - a^2}$ suggests that the distance $c = |OF_1|$ is related to the lengths a and b by the Pythagorean theorem. Figure out what right triangle to add to your sketch in order to confirm this relationship among a, b, c. After adding it, use **Measure | Calculate** to confirm that $b^2 = a^2 + c^2$.

Hyperbola in Standard Form: $\dfrac{x^2}{a^2} - \dfrac{y^2}{b^2} = 1$

Obviously, the points $(a, 0), (-a, 0)$ are on this hyperbola. Letting $c = \sqrt{a^2 + b^2}$, the foci are the points $(-c, 0)$ and $(c, 0)$. Thus, taking

$$V_1 = (-a, 0) \quad V_2 = (a, 0) \quad F_1 = (-c, 0) \quad F_2 = (c, 0)$$

we can apply **hyperbola**. The eccentricity is c/a. See Figure 3.9.

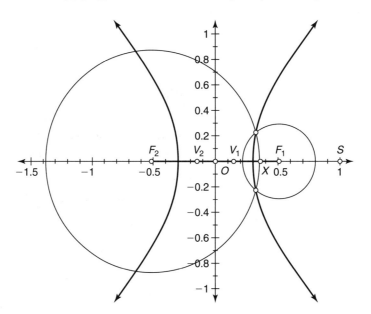

FIGURE 3.9 Hyperbola in standard position

LO2C. Carry out the construction of a hyperbola, analogous to that of an ellipse in **LO2A**. Add to your construction the asymptotes, with equations $y = bx/a$ and $y = -bx/a$. (If your tool **hyperbola** sketches only one branch, you can sketch the other branch as follows: rotate through $180°$ the point Y about the origin, and apply **Locus**.)

LO2D. The other standard form for a hyperbola is $(y/a)^2 - (x/b)^2 = 1$. In this case, the vertices and foci are on the y-axis. Carry out a construction of such a hyperbola, including asymptotes.

Parabola: $y = ax^2 + bx + c$

If $a \neq 0$, the graph of an equation of the form $y = ax^2 + bx + c$ is a parabola. Using the xy-axes provided by **Graph | Define Coordinate System**, you can quickly graph this parabola. Or, you could produce the parabola as the locus of a point controlled by a movable point on the x-axis.

We wish to enable the coefficients a, b, c to vary and to study the effects of these variations on the parabola. First, an efficient way to vary a, b, c will be presented. Then the parabola will be plotted using **Graph | Plot New Function**, so that there will be no need to produce it as a locus. However, merely graphing it will not show x, y, and (x, y) in motion as **Construct | Locus** would, so as a final step, we'll simply add these features to the graph—and while we're at it, we'll sketch the line tangent to the parabola at (x, y).

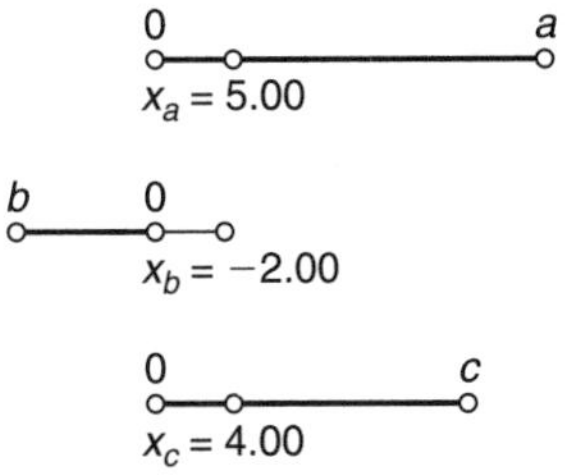

The coefficients a, b, c cannot be "input" directly as numbers. Instead, you can sketch sliders as in Section 4 of Chapter 2. If you have the tool **three sliders** handy, it can be applied here; otherwise, follow the steps in Chapter 2 to create such a tool.

Once sliders are present, the next step, as in Figure 3.10, is to add a point near the middle of the screen, select it, and apply **Graph | Define Origin**. If you wish, you can also move the expressions for x_a, x_b, x_c to the upper left corner, as shown in Figure 3.11.

FIGURE 3.10 Sliders a, b, c

$x_a = 5.00$

$x_b = -2.00$

$x_c = 4.00$

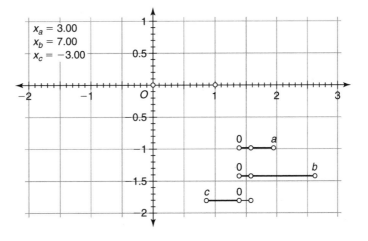

FIGURE 3.11 Abscissas. The abscissas x_a, x_b, x_c from sliders for a, b, c appear in the upper left corner, ready to be copied into the dialog box of **Graph | New Function**.

Now, to graph the parabola, apply **Graph | New Function**, using the dialog box to enter the expression

$$x_a * x\char`^2 + x_b * x + x_c$$

Not every symbol in the expression can be typed. Instead, start by clicking the expression on the main screen for x_a; in Figure 3.11, it's the equation $x_a = 3.00$. Then type the asterisk (*) or else click the asterisk on the little keyboard in the dialog box. Continue in this manner, noting that the expressions for x_b and x_c are entered in the same manner as was x_a. Then click **OK**. Check that the resulting printed expression is selected, and apply **Graph | Plot New Function**. Figure 3.12 shows a parabola graphed in this manner.

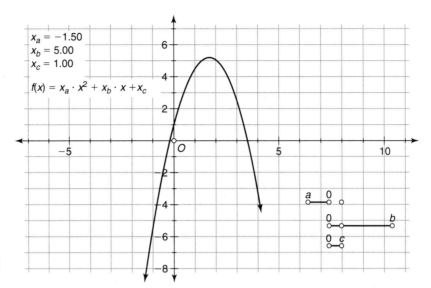

FIGURE 3.12 Parabola grapher. The graph of $y = ax^2 + bx + c$. Use sliders to vary a, b, c.

This method of graphing is quick and useful. On the other hand, there are some limitations; for example, although you can put a point P on the graph and drag it, you can't use it as a drive-point for a locus. Also, you can't use it to plot the line tangent to your parabola at P. But there is a way!

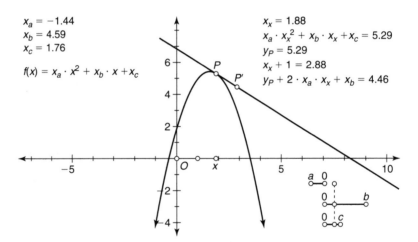

FIGURE 3.13 Parabola and tangent line. The graph of $y = ax^2 + bx + c$ with tangent line $y = 2ax + b$. Use sliders to vary a, b, c. Drag x.

A glance at Figure 3.13 suggests one way to sketch a movable tangent line. Start with an arbitrary point on the x-axis, and apply **Measure | Abscissa** to print the numerical value as x_x. Use that value with **Measure | Calculate** to compute the y-coordinate of the point on the parabola whose abscissa is x_x. Label the point P and apply **Measure | Ordinate** to print its y-coordinate. Then use **Measure | Calculate** to print the numbers

$$x_x + 1 \quad \text{and} \quad y_P + 2 * x_a * x_x + x_b$$

Click these two, in order, and apply **Graph | Plot as (x,y)**. Label the resulting point P'. Its coordinates were planned so that the slope of the line PP' is

$2 * x_a * x_x + x_b$, which is the slope of the tangent line at P. Construct line PP'. Then select x and P, and apply **Construct | Locus**. The result should resemble Figure 3.13.

ASSIGNMENT 2.3

LO2E. Create **parabola grapher.gsp** as in Figure 3.12 (or Figure 3.13). Use it to graph these curves:

$$y = x^2 + 2x + 1 \quad y = 2x + 1 \quad y = -x^2 + x + 2$$

In a caption, tell how to use this sketch to graph a line $y = mx + b$. In another caption, tell necessary and sufficient conditions for the parabola to open downward from a point in Quadrant II.

LO2F. Given 2 points on the x-axis and a horizontal line H other than the x-axis, there is a unique parabola of the form $y = ax^2 + bx + c$ that passes through the two points and has vertex on H. Construct a parabola grapher in which the user can drag the two points and line H, and the parabola will "follow".

LO2G. Given 2 points on the x-axis, one point on the y-axis and not in a vertical line with either of the first two points, there is a unique parabola of the form $y = ax^2 + bx + c$ that passes through the 3 points. Construct a parabola grapher in which the user can drag the 3 points, and the parabola will "follow".

LO2H. Construct a parabola grapher that uses six sliders to graph two parabolas,

$$y = ax^2 + bx + c \quad \text{and} \quad y = dx^2 + ex + f$$

on a single coordinate system. Figure out conditions on a, b, c, d, e, f for the parabolas to intersect in a right angle. Use **Measure** to confirm or refute your figuring.

SECTION 3   # Inversion in a Circle

Suppose $\circ(O, r)$ is a circle and P is a point other than the center O. The *inverse of P in* $\circ(O, r)$ is the point Q on ray OP that satisfies $|OP| \cdot |OQ| = r^2$. If P traverses a curve C, the locus of the inverse of P is called the *inverse of C*. Among the remarkable facts about inversion is that the inverses of lines and circles are lines and circles. Such inversions lend themselves especially well to Sketchpad constructions.

Inverse of a Point

We'll begin with the inverse of a single point P inside or on $\circ(O, r)$. Let L be the line through P perpendicular to the line OP, and let T be the line tangent to the circle at a point where L meets the circle. Then the inverse of P in $\circ(O, r)$ is the point $Q = T \cap OP$.

The inverse of Q just constructed is P, so to construct the inverse of a point P situated outside $\circ(O, r)$, we can reverse the steps in the previous construction. Specifically, let P' be a point where $\circ(O, r)$ meets the circle having segment OP as diameter (so that line PP' is tangent to $\circ(O, r)$). Let U be the line through P' perpendicular to OP. Then $Q = U \cap OP$.

ASSIGNMENT 3.1 **LO3A.** Start with independent points O, A, P. Construct $\circ(O, r)$ passing through A, so that $r = |OA|$. If P lies outside your circle, then drag P to the inside. Follow the steps printed above to construct the inverse, Q, of P. See Figure 3.14. Use **Measure | Calculate** to print both r^2 and $|OP| \cdot |OQ|$.

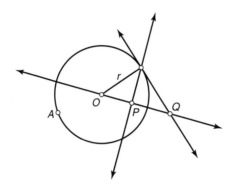

FIGURE 3.14 Inverse Q of a point P in a circle $\circ(O, r)$

LO3B. Start as in **LO3A**, but put P outside $\circ(O, r)$, and construct the inverse, Q, of P. (Consider the circle having diameter OP.) Use **Measure | Calculate** to print $|OP| \cdot |OQ|$ and r^2; be sure that these two stay equal as you vary P.

LO3C. Create a sketch for inverses in both of the cases treated in **LO3A** and **LO3B**. (Hint: Use **Measure | Calculate**, **Transform | Mark Ratio**, and **Transform | Dilate**.) Save your sketch and a tool named **inverse**.

Inverse of a Line

It is wise to regard inversion in a circle as a transformation of the whole plane (including the center O of the circle and a point at infinity serving as the inverse of O). Any figure in the plane can be inverted, and we are particularly interested in the inverses of curves. For example, you could write your name and then invert it under various circles. It would be easier, however, to start by inverting lines.

A line L passing through the center of $\circ(O, r)$ inverts to itself. This is clear from the definition of inverse of a point—just apply the definition to any point on the line (including O). If L does not pass through O, then inversion carries L onto a curve that does not contain the point at infinity. That is to say, the image-curve is bounded. In fact, it is a circle, as proved in virtually every textbook that covers the topic of inversion.

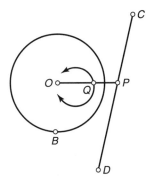

FIGURE 3.15 The smaller circle is the inverse of line *CD* with respect to the larger circle

LO3D. Start with a circle $\circ(O, r)$. Let L be a line and P a movable point on L. Use **inverse** to sketch the inverse, Q, of P in $\circ(O, r)$. Select P and Q and apply **Construct | Locus** to sketch the inverse of L, as in Figure 3.15. Drag L. In captions, print what happens when L passes through O, when L is tangent to $\circ(O, r)$, and how inverses of lines that cross $\circ(O, r)$ differ from images of lines that miss $\circ(O, r)$.

LO3E. Continuing **LO3D**, let L' be an additional line. If L' is parallel to L, they meet only at infinity, so that the inverses of L and L' meet only at O. Experiment with L and L' until you are convinced that if they are perpendicular, then their inverses are orthogonal, in the sense that at each point of intersection the angle between the two tangent lines is $90°$.

LO3F. Continuing, arrange for L and L' to avoid passing through O and for them to be not perpendicular. Measure the acute angle θ between L and L'. Let $inv(L)$ denote the inverse of L and $inv(L')$ that of L'. Figure out how to construct the centers of the circles $inv(L)$ and $inv(L')$. Then construct tangent lines to these circles at one of their points of intersection. Then measure the acute angle between the tangent lines, and discover its relation to θ.

Inverse of a Circle

Since inversion carries lines onto circles, it is natural to ask what inversion does to circles. Our recent work shows that *some* circles invert to lines. We'll confirm with Sketchpad that inversion carries *every* circle onto a line or circle. (As already mentioned, we can simplify the statement of this theorem by saying that *the inverse of a circle is a circle,* if we agree that a line is a circle with infinite radius.)

LO3G. Starting with circles c_1 and c_2, sketch the inverse of each circle in the other, and label them c_F and c_G. Print conditions under which c_F and c_G are concentric and conditions under which they are tangent. See Figure 3.16.

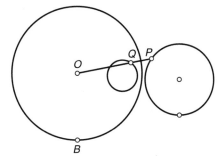

FIGURE 3.16 The smallest circle is the inverse of the outer circle with respect to the largest circle

LO3H. Start with a circle $\circ(O, r)$ and an ellipse. Sketch at least a dozen movable points on your ellipse. Invert all of them. Under what conditions do you think the inverse of an ellipse in a circle is an ellipse?

Special Conics

Among well known conics are the Steiner ellipse and the Kiepert hyperbola. In this section we'll study both of these. In particular, we'll see that they and many other conics can be obtained by taking the isogonal conjugates of the points on a line.

Suppose *ABC* is a triangle. If you apply **isogonal conjugate** (from Chapter 2) to any point on the circumcircle of $\triangle ABC$, other than *A*, *B*, or *C*, then something special happens: the three reflected lines that concur in the desired conjugate are parallel. This means that they meet at infinity. In fact, as *P* moves around the circumcircle, the directions of the parallel lines include every direction in the plane, and so the isogonal conjugate of *P* moves through the *line at infinity.* One helpful way to think about this line is simply as the set of all directions in the plane of $\triangle ABC$.

Now, suppose *L* is a line in the plane of $\triangle ABC$. When we speak of its *isogonal image,* we mean the locus of the isogonal conjugate of a point *P* that traverses the line. This image is a conic. Indeed, if *L* meets the circumcircle in two points, then the isogonal image of *L* must meet the line at infinity for two different directions, and the conic must therefore be a hyperbola. If *L* misses the circumcircle, then its isogonal image does not extend to infinity, so it must be an ellipse. This leaves the possibility that *L* is tangent to the circumcircle, and its isogonal image is a parabola.

Actually, there is a detail missing from these niceties. Recall that isogonal conjugate is undefined for points on the sidelines of $\triangle ABC$. As a point *P* traverses a line *L*, what happens when *P* takes the positions where *L* meets lines *BC*, *CA*, *AB*? For purposes of creating sketches, we needn't go into great detail. It suffices to say that the definition of isogonal conjugate can be readily extended to these positions so that the resulting conjugates are the vertices *A*, *B*, *C*. In other words, the conic sections we expect to see in sketches are *circumconics*. The prototypical circumconic is the circumcircle, which is the isogonal image of the line at infinity.

Conversely, the locus of the isogonal conjugate of a point *P* that traverses a circumconic is a line, and the line meets the circumcircle in 0, 1, or 2 points according as the circumconic is an ellipse, parabola, or hyperbola.

Steiner Ellipse

Among all the ellipses that pass through three given points *A*, *B*, *C*, the Steiner ellipse has the least area. Another important fact is that the center of this ellipse is the centroid of $\triangle ABC$.

One construction of the Steiner ellipse is as the locus of the isogonal conjugate of a point *P* that traverses a certain line. Start with $\triangle ABC$ and its symmedian point *K*, and sketch its cevian triangle, *A'B'C'*. The desired line *L* is determined by any two of the points $B'C' \cap BC$, $C'A' \cap CA$, $A'B' \cap AB$, in accord with Desargues's Theorem. See Figure 3.17.

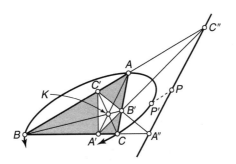

FIGURE 3.17 Triangle *ABC* and its Steiner ellipse. Drag or animate *P*.

ASSIGNMENT 4.1

LO4A. Starting with △*ABC*, construct the Steiner ellipse. Then add the centroid, *G*, of △*ABC*. Check that *G* appears at the center of the ellipse and that the isogonal conjugate of *G* is the symmedian point, *K*. You should have tools for constructing *G* and isogonal conjugates (from Chapter 2).

LO4B. Start with △*ABC* and its Steiner ellipse. Place about two dozen points on the ellipse, fairly evenly spaced. Apply **Construct | Segment** to create an inscribed polygon. Apply **Construct | Interior** and **Measure | Area** to print the area of the polygon. Drag points so that the polygon is visually nearly identical to the ellipse. Then use well-chosen points and **Measure | Distance** to estimate the lengths $2a$ and $2b$ of the major and minor axes of the ellipse. Use **Measure | Calculate** to print the product πab, a well-known formula for the area of an ellipse. Is the area of your polygon nearly equal to πab?

Two Hyperbolas

In Chapter 2, you may have noticed a certain commonality in the constructions of the circumcenter, the Fermat point, and the Napoleon point. All of these constructions involve a triangle $A'B'C'$ that is perspective to △*ABC*, and the vertices A', B', C' are on the perpendicular bisectors of sides *BC*, *CA*, *AB*, respectively. You may have suspected that there is a theorem that includes the three cases.

In pursuit of such a theorem, note something else that the three cases have in common:

$$\angle CAB' = \angle ABC' = \angle BCA'$$

This angle has measure $0°, 60°, 45°$, respectively, in the three cases cited. By taking inward-directed equilateral triangles for the Fermat and Napoleon configurations, you get two more examples, corresponding to common angles of $-60°$ and $-45°$.

To generalize, if you erect isosceles triangles on the sidelines *BC*, *CA*, *AB*, with six equal base angles adjacent to the sidelines, then the points A', B', C' determine a triangle perspective to △*ABC*. As the base angle varies from $-90°$ to $90°$, the perspector (that's $AA' \cap BB' \cap CC'$) traces a locus known as the *Kiepert hyperbola*, as in Figure 3.18.

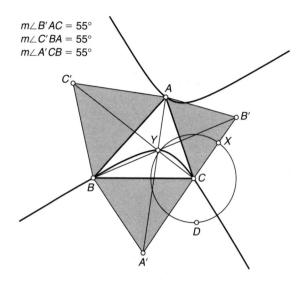

$$m\angle B'AC = 55°$$
$$m\angle C'BA = 55°$$
$$m\angle A'CB = 55°$$

FIGURE 3.18 Triangle *ABC* and its Kiepert hyperbola. Drag or animate *X*.

Another conic section that is easy to sketch is the *Jerabek hyperbola,* the locus of the isogonal conjugate of a point *P* that traverses the Euler line of △*ABC*.

LO4C. Start with △*ABC*. Following the steps printed above, use **Construct | Locus** to sketch the Kiepert hyperbola, as in Figure 3.18. To control the base angle, sketch a circle with center *C*. Put a movable point *X* on the circle, and use ∠*XCA* with **Transform | Mark Angle** and **Transform | Rotate**.

LO4D. Start with △*ABC*. Add the circumcenter, *O*, and the symmedian point, *K*. Let *X* be a movable point on line *OK*. Construct the isogonal conjugate, *Y*, of *X*, and then sketch the locus of *Y* as a function of *X*. This locus is the Kiepert hyperbola. Be sure to compare the results of **LO4C** with those of **LO4D**.

LO4E. Sketch the Jerabek hyperbola of △*ABC*, and confirm that it passes through the circumcenter, orthocenter, and symmedian point of △*ABC*.

SECTION 5 # Circles and Polygons

In this section, we'll construct one of the simplest loci of all, without using **Locus**. The locus is that of a point *P* that stays equidistant from a point *O*. Let *P* assume some initial position. Sketchpad can rotate *P* about *O* through a small angle *t* to produce a point *P'* having the same distance from *O* as *P*. Then Sketchpad can rotate *P'* through *t* to produce another point, *P''*, having the same distance from *O*. The process can be repeated to produce a large number of points on the locus—enough points to see that the locus is a circle. Of course, with only finitely many images of *P*, what we really have here are the vertices of a polygon, all of which lie on a circle. Changing t to 60° yields a regular hexagon.

Polygons

In Figure 3.19, the smaller the angle ABC, the more circular the sketch. It is worth noting, however, that if you change $\angle ABC$ to 120°, the rotated points assume only 3 positions, namely, the vertices of an equilateral triangle. Similarly, you can sketch squares, pentagons, and so on.

$m\angle ABC = 14.50°$

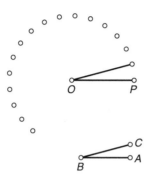

FIGURE 3.19 Locus of P equidistant from O, typified by repeatedly rotating an initial point. Drag C to vary the rotation angle. Drag O, P to vary the center and radius.

$m\angle ABC = 72.00°$

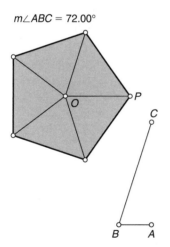

FIGURE 3.20 Pentagon sketched by rotations using $\angle ABC = 72°$

LO5A. Emulating Figure 3.19, sketch at least 50 unlabeled points on a circle. (Windows shortcut: **Alt T | R | R.**) Then sketch another circle in the same manner, having a radius that you can vary, this circle being tangent to the first with neither circle enclosing the other. Finally, use **Construct | Circle By Center+Point** to construct the smallest circle that goes around both of the other circles.

LO5B. Emulating Figure 3.19, figure out and implement an efficient way to add a radius and chord of the circle with each new rotation.

LO5C. Emulate Figure 3.19, but this time, rotate an arbitrary line instead of a point. Iterate this rotation at least 20 times.

LO5D. Repeat the steps as in **LO5C**, but this time, rotate an arbitrary circle instead of a line. Iterate this transform at least 20 times.

LO5E. Use the methods of this section to sketch on one screen regular n-gons, for $n = 3, 4, 5, 6, 8$. For each one, join the center to each vertex, and include a printed measure of one of the equal angles at the center. See Figure 3.20.

LO5F. Place a dozen or more points on the screen. Apply **Edit | Select All** and **Construct | Polygon Interior**. If any edge crosses another edge, drag vertices so that this doesn't occur. Measure the area and perimeter. Drag your points to positions for which your area and perimeter are as nearly equal as possible. The attempt to do this should enable you to make a conjecture that may be easier to "see" than to put into words. Nevertheless, in a caption, craft a careful statement of the conjecture.

PROJECTS

PROJECT 1: GENERAL CONICS

The conics given by simple forms such as

$$(x/a)^2 + (y/b)^2 = 1 \quad \text{and} \quad (x/a)^2 - (y/b)^2 = 1$$

are in standard position, but the form for the *general* conic is

$$ax^2 + bxy + cy^2 + dx + ey + f = 0$$

Writing this as

$$cy^2 + (bx + e)y + (ax^2 + dx + f) = 0$$

shows that for given x, the quadratic formula yields

$$y = (-B \pm \sqrt{B^2 - 4AC})/(2A)$$

where

$$A = c \quad B = bx + e \quad C = ax^2 + dx + f$$

Following the method for creating **LO2E** (parabola grapher, for equations of the form $y = ax^2 + bx + c$), you can use sliders for the coefficients a, b, c, d, e, f, and graph the general conic. Then by varying a, b, c, d, e, f, you can study the effects of these coefficients on the shape and the "tilt" of your conic.

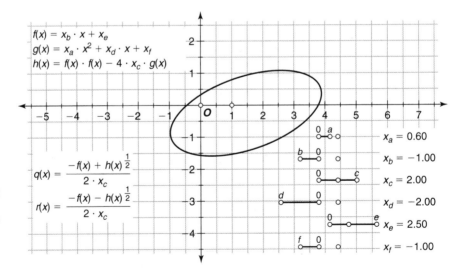

FIGURE 3.21 General conic grapher: $ax^2 + bxy + cy^2 + dx + ey + f = 0$. Use sliders to vary the coefficients a, b, c, d, e, f. Animate the sliders, especially b.

$$f(x) = x_b \cdot x + x_e$$
$$g(x) = x_a \cdot x^2 + x_d \cdot x + x_f$$
$$h(x) = f(x) \cdot f(x) - 4 \cdot x_c \cdot g(x)$$

$$q(x) = \frac{-f(x) + h(x)^{\frac{1}{2}}}{2 \cdot x_c}$$

$$r(x) = \frac{-f(x) - h(x)^{\frac{1}{2}}}{2 \cdot x_c}$$

$$x_a = 0.60$$
$$x_b = -1.00$$
$$x_c = 2.00$$
$$x_d = -2.00$$
$$x_e = 2.50$$
$$x_f = -1.00$$

General conic grapher, as shown in Figure 3.21, is a relatively sophisticated sketch. There is a *function-substitution* technique, exemplified here, that applies to many other configurations. The need for it would be clear if you attempted to type the above long expression for y directly into Sketchpad's Calculator. You'd find that there are too many symbols. To get around this, you'll need to substitute symbols as you go along. However, you can't just tell Sketchpad to "Let $B = bx + e$".

Instead, the trick is to use the fact that **Graph | New Function** lets us use "old" functions to create "new" functions. In Figure 3.21, look at the equations

$$f(x) = x_b \cdot x + x_e$$

$$g(x) = x_a \cdot x^2 + x_d \cdot x + x_f$$

$$h(x) = f(x) \cdot f(x) - 4 \cdot x_c \cdot g(x)$$

Since the formula for $h(x)$ uses both $f(x)$ and $g(x)$, the formula is much shorter and easier to work with than would otherwise be the case.

Here is a final note: the argument of the square root function, indicated in Figure 3.21 by $(\ )^{1/2}$, may be negative for some values of x. Sketchpad simply skips across these values, so that you need not do anything to avoid them.

PROJECT 2: POLYNOMIALS

Strictly speaking, a function f is a set of ordered pairs (x, y), where x belongs to a prescribed or tacitly understood set, called the *domain* of f, and no two of the ordered pairs have equal first components. We write $y = f(x)$ to abbreviate the assignment of y to x in the set f. (Sometimes, attempts are made to define *function* as a "rule"; however, many functions don't have a "rule".)

A *graph* of f is a picture representing the ordered pairs that comprise f. Accordingly, we may think of the graph as the locus of the point (x, y) where $y = f(x)$ and x traverses the domain of f. As mentioned earlier, one can use **Locus** to produce graphs of functions, but often, it's more efficient to use Sketchpad's offerings in the **Measure** and **Graph** menus.

Project 2 depends on those offerings for the purpose of graphing particularly important functions called polynomials. You are already familiar with some or all of the classes of polynomials as listed here:

Polynomials of degree 0: $f(x) = c$ (constant functions)

Polynomials of degree 1: $f(x) = mx + b$ (nonvertical lines)

Polynomials of degree 2: $f(x) = ax^2 + bx + c$ (parabolas that open vertically)

Polynomials, general: $f(x) = c_n x^n + c_{n-1} x^{n-1} + \cdots + c_1 x + c_0$

Part 1. After reviewing Figure 3.13 and **L02E–L02G**, construct a cubic grapher for

$$y = ax^3 + bx^2 + cx + d$$

where the coefficients a, b, c, d are controlled by sliders.

Part 2. Construct **cubic grapher from roots** for

$$y = a(x - r_1)(x - r_2)(x - r_3)$$

where the coefficient a is controlled by a slider and the roots r_1, r_2, r_3 can be dragged. See Figure 3.22.

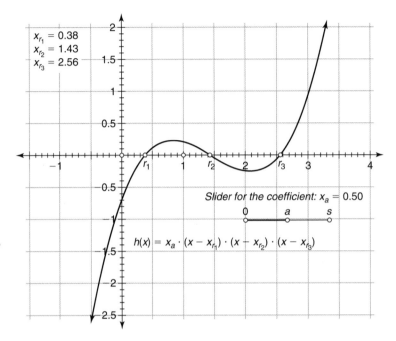

FIGURE 3.22 Cubic grapher from roots. This sketch graphs the function h formulated above. Drag the three roots on the x-axis. Use the slider to vary the coefficient a.

Part 3. Construct a quartic grapher for

$$y = a(x - r_1)(x - r_2)(x - r_3)(x - r_4)$$

where the coefficient a is controlled by a slider and the roots r_1, r_2, r_3, r_4 can be dragged.

PROJECT 3: MORE POLYNOMIALS

Loosely speaking, the higher the degree of a polynomial, the more elaborate its graph. For example, a 2nd-degree polynomial can have only one maximum or minimum point, whereas an nth-degree polynomial can have up to $n - 1$ such points. This flexibility leads to the possibility of approximating nonpolynomial functions with polynomials. For example, how "close" can a polynomial graph come to the graph of $y = \cos x$ for x in the interval $[0, 2\pi]$? For Sketchpad purposes, it would be very nice if one could select a few points on the cosine graph and then see the graph of a polynomial that passes through those same points. The objective of this project is to be able to do just that, not only for cosine, but for any reasonable function.

We begin with three points. Imagine that the user has put three points, (a, b), (c, d), (e, f) on the screen, and a sketch that you are to create will graph a parabola that passes through the three points. Then the user can then drag the points, and your parabola will "follow". What you need, of course, is a formula for an appropriate polynomial. Here it is:

$$\frac{b(x - c)(x - e)}{(a - c)(a - e)} + \frac{d(x - a)(x - e)}{(c - a)(c - e)} + \frac{f(x - a)(x - c)}{(e - a)(e - c)}$$

Obviously, when you substitute a for x in this formula, you get b, as desired, and similarly for the points (c, d) and (e, f). This clever formula was discovered by Joseph Lagrange, and the polynomials are called Lagrange polynomials. When you catch on to the "pattern" of the formula for the 2nd-degree Lagrange polynomial, just above, you can easily write down the formula for the 3rd-degree Lagrange polynomial, typified by Figure 3.23.

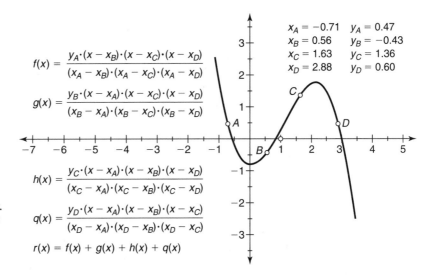

$$f(x) = \frac{y_A \cdot (x - x_B) \cdot (x - x_C) \cdot (x - x_D)}{(x_A - x_B) \cdot (x_A - x_C) \cdot (x_A - x_D)}$$

$$g(x) = \frac{y_B \cdot (x - x_A) \cdot (x - x_C) \cdot (x - x_D)}{(x_B - x_A) \cdot (x_B - x_C) \cdot (x_B - x_D)}$$

$$h(x) = \frac{y_C \cdot (x - x_A) \cdot (x - x_B) \cdot (x - x_D)}{(x_C - x_A) \cdot (x_C - x_B) \cdot (x_C - x_D)}$$

$$q(x) = \frac{y_D \cdot (x - x_A) \cdot (x - x_B) \cdot (x - x_C)}{(x_D - x_A) \cdot (x_D - x_B) \cdot (x_D - x_C)}$$

$$r(x) = f(x) + g(x) + h(x) + q(x)$$

$x_A = -0.71 \quad y_A = 0.47$
$x_B = 0.56 \quad y_B = -0.43$
$x_C = 1.63 \quad y_C = 1.36$
$x_D = 2.88 \quad y_D = 0.60$

FIGURE 3.23 Lagrange four. Graph of the simplest cubic polynomial that passes through points A, B, C, D. Drag these points.

Part 1. Create **Lagrange three** to pass through three points A, B, C as described above. Predict and confirm what happens when A, B, C are on the line $y = 2x$. Then predict and confirm positions for A, B, C for which the graph coincides with the polynomial $y = x^2$.

Part 2. Create **Lagrange four** to pass through four points A, B, C, D as typified by Figure 3.23. Predict and confirm what happens when $A = (1, 2)$ and $B = (1, 3)$.

A curve (or a pair of points) Γ is said to be *symmetric* (*about the origin*) if for every point (x, y) on Γ, the point $(-x, -y)$ is on Γ. Suppose that A and B are symmetric and that C and D are symmetric. Do you think this will force the Lagrange polynomial of A, B, C, D to be symmetric? Experiment with **Lagrange four** until you are fairly sure of the answer. Can you prove your answer?

Part 3. Starting with **Lagrange four**, add a graph of the circle $x^2 + y^2 = 9$. Predict where to put A, B, C, D so that the polynomial will approximate the top half of your circle. Then drag A, B, C, D to those positions and see how close the approximation is. Should it matter if you reverse the positions of

A and *B*? (Of course, a sketch appropriately named **Lagrange six** could approximate the semicircle even more closely than **Lagrange four** can.)

Part 4. A curve Γ is said to be *symmetric* (*about the y-axis*) if for every point (x, y) on Γ, the point $(-x, y)$ is on Γ. Suppose that *A* and *B* are symmetric and that *C* and *D* are symmetric. Will this force the Lagrange polynomial of *A*, *B*, *C*, *D* to be symmetric? Experiment with **Lagrange four** until you are fairly sure of the answer. Can you prove it?

PROJECT 4: FOCI AND A PASS-THROUGH POINT

Part 1. Let F_1, F_2, and *P* be independent points. Sketch the ellipse that passes through *P* and has foci F_1 and F_2. Sketch the extremities of the major and minor axes of your ellipse, and label them V_1, V_2, V_3, V_4. Save a tool named **ellipseFFP** for which the givens are F_1, F_2, *P* and the results are V_1, V_2, V_3, V_4 and the ellipse.

Part 2. Let F_1, F_2, and *P* be independent points. Sketch the hyperbola that passes through *P* and has foci F_1 and F_2. Sketch the vertices of your hyperbola, and label them V_1 and V_2. Save a tool named **hyperbolaFFP** for which the givens are F_1, F_2, *P* and the results are V_1, V_2, and the hyperbola.

Part 3. Let F_1 and F_2, be independent points. Let c_1 be a circle, and let *P* be a movable point on c_1. Use tools from **Parts 1** and **2** to sketch both the ellipse and the hyperbola that have foci F_1 and F_2 and pass through *P*. Animate *P*.

Part 4. Returning to the ellipse in **Part 1**, sketch the line tangent to the ellipse at a movable point *Q* on the ellipse.

PROJECT 5: INVERSIONS

Part 1. Confirm that the circle $x^2 + y^2 = r^2$ inverts a point (s, t) to the point

$$\left(\frac{r^2 s}{s^2 + t^2}, \frac{r^2 t}{s^2 + t^2} \right)$$

Part 2. Confirm visually and prove analytically that the inversion in **Part 1** carries a noncircular ellipse in standard position into a curve that isn't an ellipse.

Part 3. Create a tool named **inverse of line** for which the givens are the center of a circle c_1 of inversion, the circle itself, and an independent line, *L*, and the result is the inverse of *L* in c_1. (The inverse is a circle that you should construct without using **Locus**; that way, when you apply the tool, you'll always get a *complete* circle.)

Part 4. Sketch a circle $\circ(O, |OZ|)$ and color its interior lightly. Sketch a square grid consisting of 9 horizontal lines and 9 vertical lines. Apply **inverse of line** to each of your 18 lines. The result should be 18 circles passing through *O*. (Each pair of them are either tangent or else meet at 90°.)

Part 5. Continuing, merge point *O* to a circle centered at the middle point of your grid. Animate *O*, slowly, to witness some amazing action.

CHAPTER FOUR

Animation

SUPPOSE YOU HAVE a sketch in which a movable point P has been placed on an object U, such as a line or a circle. By selecting P and applying **Display | Animate**, you'll see P move on U, and all the children of P and U will move accordingly. For example, if U is a circle, and you've constructed the line T tangent to U at P, when you animate P, the line T as well as P will rotate around U, along with P.

SECTION 1   Area of a Triangle

A common formula for the area of a triangle is $A = bh/2$, where b is a base and h is the height of the altitude from that base. Any one of the sides can be the base, and this suggests a sketch that calculates the area in three different ways. See the printed measurements of areas in Figure 4.1.

One-half Base Times Height

For a given triangle UVW with base VW of length $b = |VW|$ and height $h =$ distance from vertex U to line VW, you can easily sketch other triangles

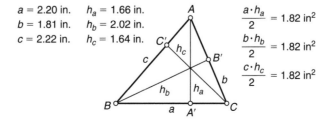

$a = 2.20$ in. $h_a = 1.66$ in.
$b = 1.81$ in. $h_b = 2.02$ in.
$c = 2.22$ in. $h_c = 1.64$ in.

$$\frac{a \cdot h_a}{2} = 1.82 \text{ in}^2$$

$$\frac{b \cdot h_b}{2} = 1.82 \text{ in}^2$$

$$\frac{c \cdot h_c}{2} = 1.82 \text{ in}^2$$

FIGURE 4.1 Area = one-half base times height, three ways

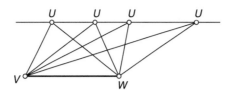

FIGURE 4.2 Triangles *UVW* having the same area

having the same base and height: just keep *V* and *W* where they are and move *U* so that it stays a fixed distance from line *VW*, as in Figure 4.2.

AN1A. Emulate Figure 4.1, including the printed measurements. Select point *A* and apply **Display | Animate**. (Since *A* is an independent point, its motion under animation is "random".) Practice changing the speed, pausing, resuming, and apply **Display | Stop Animation**. Then select all three points, *A*, *B*, *C* and apply **Display | Animate**. As always, print your observations.

AN1B. Start with independent points *V* and *W*, a line *L* parallel to line *VW*, and a movable point *U* on *L*. Then select point *U* and apply **Display | Animate**. As *U* moves, its distance *h* from line *VW* stays constant. Also, the base-segment *VW* stays fixed. Therefore, the area of the changing triangle *UVW* stays constant. Halt the animation and use **Measure** to print the area of △*UVW*, and use **Measure | Calculate** to print $h \cdot |VW|/2$. Animate *U* again, and observe that this calculated area stays constant.

AN1C. Add to **AN1B** a printed calculation of one-half base times height using segment *WU* as base. Note that this calculation agrees with the earlier one; here, however, the length of the base varies.

Heron's Formula

Suppose you are given only the sidelengths, *a*, *b*, *c* of a triangle. The three lengths determine a triangle, so there must be some formula for the area in terms of *a*, *b*, *c*. It is known as Heron's formula, after Heron of Alexandria—although there is evidence that Archimedes knew this formula before Heron's time.

Heron's formula may seem surprisingly elaborate when written out in full. Here it is:

$$\text{area}(\triangle ABC) = (1/4)\sqrt{(a+b+c)(-a+b+c)(a-b+c)(a+b-c)}$$

How many ways can you relabel the three sides, using the letters *a*, *b*, *c*? For example, you could leave *a* where it is and interchange *b* and *c*. If you do this, you haven't changed the area of the triangle—so Heron's formula should give the same answer. There are actually six different ways to label the sides using *a*, *b*, *c*. Convince yourself that Heron's formula gives the same answer for all six ways.

AN1D. Start with a triangle as in **AN1A**. Print the area of $\triangle ABC$, calculated by Heron's formula, right next to the earlier printed area calculated the other way. Be sure the calculations agree when you animate A.

AN1E. Start with a circle and chord BC. Place a movable point A on the circle and sketch $\triangle ABC$. Apply **Measure | Area** to print the area of $\triangle ABC$. Then calculate the area of $\triangle ABC$ using Heron's formula. Animate A. In contrast to the animation in **AN1A** here, the point A is not independent, and the animation is much more predictable.

SECTION 2 Angles and Orientation

Under **Edit | Preferences | Angle**, Sketchpad gives you a choice between measuring in "degrees" and "directed degrees". This choice prompts the topic of orientation. By convention, counterclockwise is the mathematically positive orientation. The alternative is negative orientation, or clockwise. Thus, when measuring angles with "directed degrees", you get $\angle ABC = -\angle CBA$; but with "degrees", you get $\angle ABC = \angle CBA$. In this section, be sure to use "degrees", not "directed degrees".

Central Angle in a Circle

One of the most useful theorems about angles is the central-angle theorem, illustrated by Figure 4.3 and stated as follows. *Suppose $\circ(O, |OA|)$ passes through points A, B, P and $\angle APB < 180°$. Then*

$$\angle AOB = \begin{cases} 2 \cdot \angle APB & \text{if } P \text{ lies on the longer of the 2 arcs from} \\ & A \text{ to } B \\ 360° - 2 \cdot \angle APB & \text{otherwise} \end{cases}$$

A proof using isosceles triangles OBP and OPA follows. If P lies on the longer arc from A to B, then

$$\angle AOB = 360° - \angle BOP - \angle POA$$

$$= 360° - (180° - \angle PBO - \angle OPB) - (180° - \angle OAP - \angle APO)$$

$$= \angle PBO + \angle OPB + \angle OAP + \angle APO$$

$$= 2(\angle OPB + \angle APO)$$

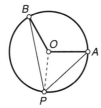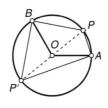

FIGURE 4.3 Central angle theorem: two cases

as required. On the other hand, if P lies on the shorter arc from A to B, we use the right angles PAP' and PBP' (in Figure 4.3) to write

$$\angle AOB = 2 \cdot \angle AP'B$$

$$= 2(90° - \angle APP' + 90° - \angle BPP')$$

$$= 360° - 2 \cdot \angle APB$$

Eliminating $\angle POA$ leaves

$$\angle AOB = 2(\angle OPA - \angle OPB)$$

as required.

ASSIGNMENT 2.1

AN2A. Sketch a circle $\circ(O, |OZ|)$ with chord AB. Place a movable point P on the circle and sketch segments AO, BO, PA, PB. Measure angles as needed to confirm the central angle theorem. Animate point P. When is the angle at point P a right angle?

AN2B. Add to **AN2A** printed measures of the areas of triangles OAB and PAB. In how many positions of P on the circle will the two triangles have equal areas? In how many positions will one of the areas be twice the other? Add the midpoint of segment AP, and predict and sketch its locus.

Forwards and Backwards at the Same Time

In this section, we are interested in animating a point P around a circle $\circ(O, |OA|)$ while a point Q goes around another circle, $\circ(U, |UB|)$. We want P to go counterclockwise while Q goes clockwise. Using your knowledge of Sketchpad, and using pencil and paper, try to figure out how to sketch P and Q. If successful, or after a ten minutes of trying, go on to the next paragraph.

One way to accomplish this is to reflect P about a diameter of its circle, obtaining a point P' whose motion is clockwise. Then P' can be translated by vector OU to a point P'' that goes around $\circ(U, |UA|)$. If $|UB| = |UA|$, we're finished. Otherwise, there are several easy ways to construct the desired point Q from P''.

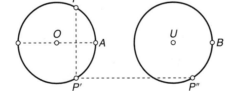

FIGURE 4.4 When point P rotates, its children P' and P'' rotate the other way

ASSIGNMENT 2.2

AN2C. Start with $\circ(O, |OA|)$ and $\circ(U, |UB|)$, where B is not a child of A. Place a movable point P on the first circle, and construct Q on the second as a child of P, following the procedure given above, so that when P is animated, point Q will traverse its circle with orientation opposite that of P.

AN2D. Add to **AN2C** the following:

tangent lines at points P and P'',

the point R of intersection of the tangent lines.

Select P and R and prepare to apply **Construct | Locus**. Before you do, try to predict whatever you can about the resulting locus of R. Finally, animate P.

SECTION 3 # Discretization

It is oversimplified but helpful to say that there are two kinds of mathematics: continuous and discrete. (Mathematics that combine both has been called "concrete"; that's con+crete from *con*tinuous and dis*crete*.) Simple animation typifies continuous mathematics since it occurs in the manner of a variable moving smoothly along a line or around a circle. By contrast, discrete mathematics has variables that "jump" from one value to another—like counting from 1 to 10.

In this section, we'll see how to convert continuous action into discrete action. The continuous action will be motion of a point going around a circle. We'll first convert this motion into a measurement that snaps from 0 to 1 to 2 to 3 and then repeats this cycle. Once we have a measurement that behaves discretely like this, we'll harness it to drive discrete geometric actions.

Snapping from 0 to 1 to 2 to 3

You may have noticed, among the functions listed as available on Sketchpad's **Calculator**, the truncation function, abbreviated as **trunc**. This function is perhaps already familiar to you by one of its more common names, such as "floor function" and "greatest integer function". The latter, especially, helps one remember the definition:

$$\text{trunc}(x) = \text{greatest integer} \leq x$$

For example, $\text{trunc}(\pi) = 3$, and $\text{trunc}(2) = 2$, as indicated by Figure 4.5. Simplify this: $\text{trunc}(-\pi)$.

Suppose you have points P and Q on a circle with center O. Suppose further that P stays fixed while Q moves around the circle. With "directed degrees" chosen for the measure of $\angle POQ$, the range of values of this measure is from $-180°$ to $180°$. This measure can be regarded as a continuous variable, and we wish to use it with trunc to create a discrete variable. First, note that $\angle POQ + 180°$ varies from $0°$ to $360°$. If $\angle POQ + 180$ is divided by 90, the resulting variable ranges continuously from 0 to 4. Thus trunc, applied to $(\angle POQ + 180)/90$, leaves a discrete variable that takes only the values 0, 1, 2, 3.

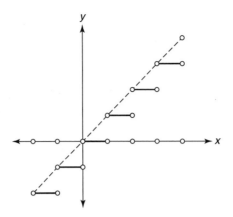

FIGURE 4.5 Graph of
$y = \text{trunc}(x)$

ASSIGNMENT 3.1

AN3A. Follow the method described above to print the calculated value of $\text{trunc}(\angle POQ + 180)/90$, and be sure that this value repeatedly cycles through the numbers 0, 1, 2, 3.

AN3B. Starting with **AN3A**, use the cycling measurement (call it m) to sketch a circle whose radius is m. When $m = 0$, your circle should be just a point; when m snaps to 1, your circle should suddenly take a radius of 1, and so on.

Snapping from 0 to 1 to ... M

Next, change 90 to N, so that $(\angle POQ + 180)/N$ will range from 0 to $360/N$, and $\text{trunc}((\angle POQ + 180)/N)$ will snap through the values $0, 1, 2, \ldots, M$, where $M = \text{trunc}(360/N) - 1$.

If $N < 90$, we can expect the speed of the "discrete animation" to exceed the speed when $N = 90$. In fact, if $N = 1$, the speed should be the same as if there were no discretization at all, and the motion would appear to be continuous.

ASSIGNMENT 3.2

AN3C. Experiment with **AN3B**, for which $N = 90$, by taking N to be 60, then 30, and then 12.

AN3D. Continuing, arrange for one of the M circles to have radius $1/2$ and center $(1/2, 0)$; another, radius $2/2$ and center $(2/2, 0)$; another, radius $3/2$ and center $(3/2, 0)$; and so on.

Randomization

At present there is no generally accepted definition of *random sequence*. The best that computers can do is to generate "pseudorandom numbers". However, on Sketchpad, there seems to be no fast "automatic" way to access such numbers. In this section, we'll generate our own pseudorandom numbers using some relatively patternless function of a point r traversing a horizontal segment.

Of course, 9^r is not "random", but its fractional part is sufficiently unpredictable for our purposes if we keep $r > 1$. A formula for this fractional part is

$$f(r) = 9^r - \text{trunc}(9^r)$$

The numbers $f(r)$ are between 0 and 1. Should you wish to have pseudo-random *integers* from 0 through 9, you can use the function

$$g(r) = \text{trunc}(10 * f(r))$$

Or, if you need pseudorandom 0's and 1's, use the function

$$h(r) = \text{trunc}(2 * f(r))$$

ASSIGNMENT 3.3

AN3E. On a square grid, let $U = (1, 0)$, and let V be a movable point on the x-axis, a few inches to the right of U. Print the abscissa x_r of a movable point r on segment UV. Use **Measure | Calculate** to print $f(x_r)$ as formulated above. Select x_r and $f(x_r)$ and apply **Graph | Plot As (x, y)**. Label the resulting point H. Sketch segment rH. Select it and apply **Display | Trace Segment**. Then animate r. Your result should look "random". Replace $f(r)$ by other functions that you think might result in segments rH that look "random".

AN3F. Create a sketch in which two disks as in Figure 4.6 change radii randomly. The radius of one disk should take the values 1 and 2 as determined by suitably modified values of $f(x_r)$ as in **AN3E**. Vary the radius of the other disk as you wish.

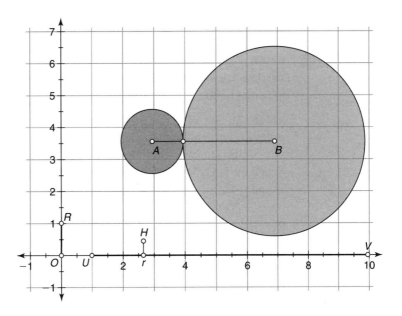

FIGURE 4.6 Random disks

Transformations

Sketchpad's **Transform** enables you to carry out certain basic transformations of geometry, called *translation, rotation, dilation,* and *reflection.* Each of these applies to a point, and therefore, each applies to any set of points, including Sketchpad objects. We'll start with **Translate**, which shifts selected objects a specified distance in a specified direction. That is, a translation is determined by a vector, which in turn is determined by an initial point (or tail) and a terminal point (or head).

Translation

A translation is determined on Sketchpad when you select two points and apply **Transform | Mark Vector**. Then, select any collection of objects to be translated, and apply **Transform | Translate**. A dialog box will remind you that a vector has been marked. (The box also enables other ways to decree a vector to determine the translation.)

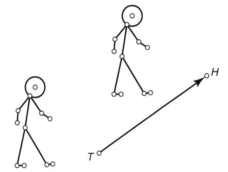

FIGURE 4.7 Humanoid, translated by vector *TH*

AN4A. Choose **Graph | Define Coordinate System**, sketch an object and then use **Translate** to move your object to the right 1 unit and up 2 units. Do this first in two separate translations. Then do it in one translation.

AN4B. Let *AB* be a line and *C* a point not on *AB*. Let *P* be a movable point on *AB*, and sketch segment *CP*. Construct an object and call it *O*; this can be simply a point or something elaborate, but you'll probably want to keep it fairly small. Use **Transform** to translate *O* by vector *CP*. Then animate *P*.

Rotation

A rotation is determined by a point, called the center of rotation, and a directed angle of rotation. As we've already seen, it is easy to mark angles for rotation by placing a movable point *P* on $\circ(O, |OZ|)$. The point *P* can be moved to produce the desired $\angle ZOP$, which can be marked for use in **Transform | Rotate**. After marking an angle, select a point and apply **Transform |**

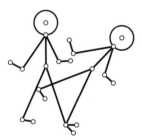

FIGURE 4.8 Humanoid, rotated 45° clockwise about right heel

Mark Center. Then apply **Transform | Rotate.** Dramatic results can be obtained in this way when the angle of rotation varies as part of a construction.

ASSIGNMENT 4.2

AN4C. Create a coordinate system, sketch an object, and then rotate your object about the origin counterclockwise through an angle of 45°. Determine whether Sketchpad enables rotating the coordinate system, including the grid.

AN4D. Place a movable point P on $\circ(O, |OZ|)$, a movable point R on segment OZ, and a movable point S on $\circ(P, |OR|)$. Measure $\angle ZOP$, select the measurement, and use **Measure | Calculate** to form $2 * \angle ZOP$. Select this calculation, select point P as a center, and rotate point S about P through $2 * \angle ZOP$. Label the resulting point Q. Select P and Q and apply **Construct | Locus.** First, try to predict whatever you can about this locus. Then animate P.

Dilation

A dilation is determined by a center and a number. For example, if the center is the origin and the number is $1/2$, then each point (x, y) gets transformed to the point $(x/2, y/2)$. The number can be a typed-in scale factor or a **Marked Ratio** determined by selecting two segments. Letting n denote the length of the first segment and d the length of the second, the **Marked Ratio** is n/d.

ASSIGNMENT 4.3

AN4E. Sketch a humanoid, as in Figure 4.9. Use **Transform | Dilate** to dilate the humanoid from at least three different centers on the screen. Try several scale factors.

AN4F. Sketch a line AB, a circle O that misses the line, and a movable point P on O. Let L be the line through P perpendicular to line AB, and let $P' = L \cap AB$. Let $s = |PP'|$, and let r be the radius of O. Let Q be an independent point. Let Q' be the dilation of Q from P using scale factor s/r. Sketch the locus of Q' as a function of P. Animate P.

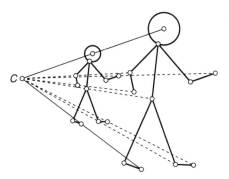

FIGURE 4.9 Humanoid, dilated about center C using scale factor 2

Reflection

A key word in connection with reflection is *mirror*. In order to reflect a selected object in a line, the line must first be selected and marked using **Transform | Mark Mirror**.

ASSIGNMENT 4.4

AN4G. Reflect a humanoid in a line. Then reflect that reflection in a second line. Now vary the lines.

AN4H. Sketch a movable point P on $\circ(O, |OZ|)$, line OP, and the line t tangent to the circle at P. Place an independent point Q not far from P. Sketch Q' as the reflection of Q about line OP, and then sketch Q'' as the reflection of Q' about line t. Show the locus of Q'' as a function of P, and animate P. As usual, make predictions.

FIGURE 4.10 Humanoid reflected in a line

Composites

Having now sampled each of the four elementary transformations—translation, rotation, dilation, and reflection—we are in a position to view these as "elements" with which to compose "compounds". For example, a compound transformation called *glide reflection* occurs when a reflection is followed by a translation in a direction parallel to the mirror.

Three of the basic transformations carry objects onto images that are congruent to the originals; in other words, these transformations preserve distance. The only basic transformation that does not preserve distance (unless the ratio used is ± 1) is dilation. The Latin roots for "same distance" being essentially "iso" and "metric", it is natural to describe as *isometric* these three transformations: translation, rotation, and reflection. An important theorem states that composites of these are also isometric. That is, if you use any number of these three elementary transformations, in any order, the resulting composite transformation is an isometry. As a converse of the theorem, every isometry is a composite of at most three reflections.

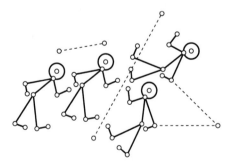

FIGURE 4.11 Humanoid translated, reflected, and rotated

AN4I. Sketch the triangle having vertices

$$A = (0, 0) \quad B = (4, 0) \quad C = (0, 3)$$

Figure out a composite of elementary transformations that carries $\triangle ABC$ onto the triangle T that has vertices $(5, 3)$, $(5, 6)$, $(9, 6)$. Use offerings in the **Transform** menu to perform this composite transformation (an isometry), and sketch the image of the point $(-1, -2)$. Then figure out an isometry that carries T onto $\triangle ABC$.

AN4J. Sketch a movable point P on a circle $\circ(O, |OZ|)$, line OP, and a line L near the circle. Sketch the reflection P' of P in line L and the reflection P'' of P' in line OP. Sketch the locus of P'' as a function of P. Animate P on its circle. Be sure to vary the location of line L.

AN4K. Create a sketch that confirms the following proposition: the composite of three reflections in three concurrent lines is a reflection in a line that passes through the point of concurrence.

AN4L. Create a sketch that confirms the following proposition: if two objects are congruent, then there exists a composite of three elementary isometries that transforms one object into the other.

Similarity

Two geometric figures are *congruent* if each can be obtained from the other by an isometry. But, as you know, sometimes figures have identical shapes without having identical sizes—then they are *similar* rather than congruent. A major theorem states that two figures are similar if one can be obtained from the other by an isometry followed by a dilation.

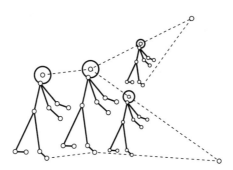

FIGURE 4.12 Similar humanoids

AN4M. Sketch the triangle having vertices

$$A = (0, 0) \quad B = (4, 0) \quad C = (0, 3)$$

Figure out a composite of elementary transformations that carries $\triangle ABC$ onto the triangle having vertices $(5, 4)$, $(5, 6)$, $(23/3, 6)$. Use options under **Transform** to perform this same composite transformation (a similarity) on the triangle having vertices $(-1, -2)$, $(0, 0)$, and $(2, -2)$.

AN4N. Sketch rays *AB* and *AC*, put a movable point *P* on ray *AB*, sketch the reflection *Q* of *P* in ray *AC*, and sketch segment *AQ*. Color the interior of △*APQ*. Drag *P* so that △*APQ* is small, and sketch a segment *DE*, and angle *UVW*, and a line *L*. Denote △*APQ*, including its vertices, edges, and interior, by *T*. Translate *T* by vector *DE*, rotate the result about a point of your choice using ∠*UVW* as angle of rotation. Then reflect the resulting image about line *L*, and dilate the result using a marked center and marked ratio of your choice. Print measurements to confirm that the final triangle is similar to △*APQ*. Then animate *P*.

SECTION 5 Speed

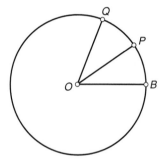

FIGURE 4.13 Point *Q* rotates twice as fast as point *P*

The notion of continuous motion prompts questions about speed. On Sketchpad, the main kind of continuous motion is invoked by **Display | Animate**, which applies to motion of a point on a circle or line, or some variant thereof. Starting with a fixed point *B* and a movable point *P* on a circle ∘(*O*, |*OZ*|), you can measure ∠*BOP*, apply a function *f* to this measurement, and use this calculation in subsequent constructing. A simple example is given by the function $f(t) = 2t$. Rotating *B* through angle 2 · ∠*BOP* gives a point *Q* whose speed is twice that of point *P*.

An analogous procedure lets you double the speed of a point *P* that moves on a line. Of course, for both circular and linear motion, you can use other functions *f* to achieve triple speed, half speed, and so on.

ASSIGNMENT 5.1

AN5A. Construct an animated configuration in which a point *Q* moves around a circle twice as fast as a point *P*. Add tangent lines at both points, and plot the locus of the point of intersection of the two tangent lines. Animate *P*.

AN5B. Use the method of this section to sketch a clock with moving hour and minute hands. Animate the hour hand. (Create an **Animation Button** to produce *clockwise* rotation.)

SECTION 6 Rolling a Wheel

Sketchpad can mimic the rolling of a circle on a line and thereby show an interesting curve called a *cycloid*. The main idea underlying this mimicry is to compose two transformations—specifically, to perform translation and rotation simultaneously. The effect is indicated by Figure 4.14; just imagine the circle rolling along the line *UA*.

AN6A. Emulate Figure 4.14, starting with **Graph | Define Coordinate System | Hide Grid**. Let A be a point, and let UV be the line passing through A parallel to the x-axis. Let P and F be movable points, and let A and B be fixed points on a circle as shown in Figure 4.14. The purpose of point F is to enable you to measure the arclength, d, of arc(BFP). Place B' at distance d to the right of B. Now for the big step: translate the circle and all its children except F by vector $\overrightarrow{BB'}$. Label the image of P as P'. Use **Locus** to produce the cycloid. Then animate P. (Adjust the scale of the axes so that 1 radian of horizontal distance equals 1 radian of arclength.)

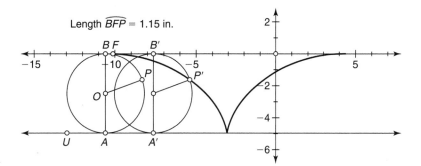

Length $\overarc{BFP}$ = 1.15 in.

FIGURE 4.14 Locus of P' on rolling circle is a cycloid

AN6B. Continuing, place a movable point inside segment OP and construct its locus as a function of P. This locus is a *trochoid*.

PROJECTS

PROJECT 1: LOCUS OF FOCUS

In this project, functions of a very interesting type will be sampled using animation. Most functions you have encountered have a set of numbers for both domain and range. Here, however, you will encounter a range that is a set of curves.

Part 1. Let X be a movable point on a circle $\circ O$. Let D be a line and X' the perpendicular projection of X on D. Sketch the reflection F of X' in X. Follow the construction for **LO1A** to obtain the parabola with focus F and directrix D. Then animate X. You will see the parabola change continuously as a function of X.

Halt the animation and construct the locus of F as a function of X. Then animate X again. This animation illustrates two functions of X. The range of one is a set of parabolas, and the range of the other is an ellipse. See Figure 4.15.

Part 2. Construct other functions of X in this manner. You might start with simple additions to the sketch for **Part 1**, but then try some other possibilities.

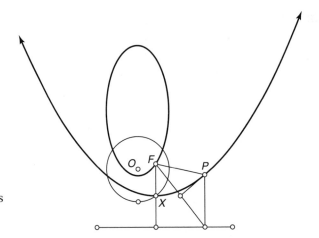

FIGURE 4.15 Two functions of X: a parabola and an ellipse

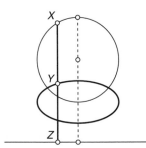

FIGURE 4.16 The locus of midpoint Y, as a function of X, is an ellipse

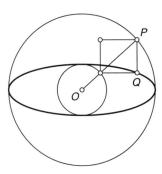

FIGURE 4.17 The locus of Q is an ellipse

PROJECT 2: INVERSE OF ANIMATED LINE

Let P be a movable point on a circle $\circ(O, |OA|)$, where O is the origin of an xy coordinate system and A is a point in Quadrant I near the point $(1, 0)$. Let D be a point on your circle, between A and P, quite close to A. Let t denote the arclength of arc ADP, and let V be the vertical line $x = t$. Invert V with respect to the circle, as in **LO3C** (or, better, using the tool **inverse of line** from Chapter 3, Project 5, Part 4). Let W be the line tangent to the circle at P, and invert W. Now animate P.

PROJECT 3: ECCENTRICITY

In Figure 4.16, imagine X going around the circle. Its child Y is the midpoint of segment XZ. If "midpoint" is changed so that the ratio $d = |XY|/|XZ|$ is a fixed number other than $1/2$, will the locus of Y be a conic? If so, figure out its eccentricity in terms of d.

PROJECT 4: ELLIPSE BETWEEN CIRCLES

Part 1. Emulate Figure 4.17. Prove analytically in a caption that the locus of Q is an ellipse.

Part 2. Let Q' denote the vertex opposite point Q in a rectangle as in Figure 4.17. Prove, in a caption, that the locus of Q' is an ellipse.

Part 3. Let R be the reflection of Q in the line OP. Sketch the locus of R as a function of P.

PROJECT 5: CONFOCAL CONICS

Use the methods of Chapter 3 to graph the equation

$$\frac{x^2}{a^2 - t} + \frac{y^2}{b^2 - t} = 1$$

where a, b, t are controlled by sliders. (See Figure 4.18.) Animate t, a, and b separately. For fixed values of a and b, what values of t yield ellipses? In a caption, print a proof that for fixed a and b, these conics have the same foci. Construct the foci.

PROJECT 6: CONICS BY "DILATING A CIRCLE"

Part 1. Create a sketch which dilates each point $P = (x, y)$ on the unit circle

$$x^2 + y^2 = 1$$

to a point $P' = (x', y')$ that lies on the ellipse

$$\frac{x^2}{a^2} + \frac{y^2}{b^2} = 1$$

where the positive numbers a and b are controlled by sliders. Sketch the locus of P' as a function of P. Animate a, b, and P, separately.

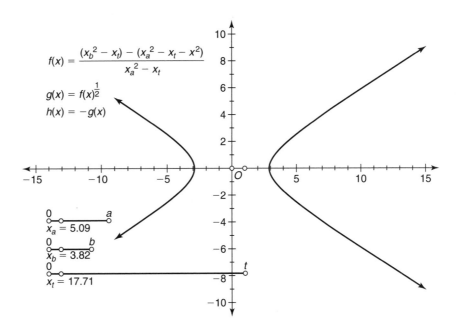

FIGURE 4.18 Confocal conics. For fixed a and b, the conics have the same foci. Animate t.

Part 2. Adapt your sketch for **Part 1** so that the locus of P' is the hyperbola

$$\frac{x^2}{a^2} - \frac{y^2}{b^2} = 1$$

PROJECT 7: ANIMATED HUMANOIDS

Design a humanoid $\mathcal{H}$ that is a child of a movable point P on a circle. Use translations, reflections, and rotations to create several clones of $\mathcal{H}$. Animate P. Experiment with the effects of **Display | Trace Objects**, using various choices of **Number of Samples** in **Advanced Preferences**.

Trigonometry

A CERTAIN GEOMETRIC curve is one of the world's most important—in fact the Latin word for *curve* became its name: *sine* (from *sinus*). Another indication of the importance of this special curve is that the entire subject called Trigonometry is the study of this curve, together with its many off-spring. Among them are *cosine, tangent, cotangent, secant, cosecant,* and many interrelationships among these curves.

In this chapter, we'll start with the definition of sine and the manner in which sine begets the other five basic trig functions. After that, we'll explore relationships among these functions.

Strictly speaking, a function is a set of ordered pairs, no two of which have the same first component. However, instead of speaking of the whole function, with notations like f or sine, we often speak only of the values of the function, with notations like $f(x)$ and $\sin t$. Just to be crystal clear about this distinction, "$\sin t$" is a value depending on t, whereas "sine" is the whole curve; that is, the following set of ordered pairs:

$$\text{sine} = \{(t, \sin t) : t \text{ is a real number}\}$$

Thus, to define sine, we must define $\sin t$ for every t. To do this, we'll use the circle $x^2 + y^2 = 1$. First, apply **Graph | Define Coordinate System**, label the origin as O and the point $(1, 0)$ as U. Select O and U in that order, and apply **Construct | Circle By Center+Point**. You now have "the unit circle". Use **Edit | Preferences** to set **Angle Units** to radians, so that both the angle and the arclength it subtends have the same measure. Place a movable point P on the circle, and measure $\angle UOP$. To keep notation simple, let $t = \angle UOP$ as in Figure 5.1. We are now ready for the fundament of this chapter:

The number $\sin t$ is, *by definition,* the y-coordinate of P.

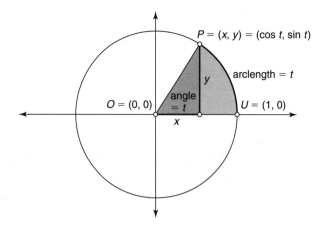

FIGURE 5.1 Birthplace of trigonometry: the unit circle, $x^2 + y^2 = 1$

Similarly, $\cos t = x$-coordinate of P. In Section 5, we'll prove that

$$\cos t = \sin(t + \pi/2) \text{ for all } t$$

This identity shows that cosine is geometrically congruent to sine, and in fact is the quarter-period left-shift of sine. In this sense (and others) cosine is an offspring of sine.

SECTION 1 $x = \cos t,\ y = \sin t$

By animating point P on the unit circle, it becomes especially clear that t is an independent variable, and that $x(t)$ and $y(t)$—that's $\cos t$ and $\sin t$—are dependent variables. As t moves smoothly around the circle, the projection (or shadow) of P on the x-axis—again, that's $\cos t$—moves periodically between -1 and 1. Likewise, the projection of P on the y-axis moves between -1 and 1. In both cases, the motion is fastest at 0 and slowest at ± 1. When one of the two functions takes the value 0, the other takes its maximal or minimal value.

ASSIGNMENT 1.1

TR1A. Emulate Figure 5.1, and animate P. As always, print your observations.

TR1B. Continuing, determine all t satisfying $|\sin t| = |\cos t|$.

TR1C. Continuing from **TR1A**, emulate Figure 5.2, obtaining one cycle of sine as shown by using **Locus** (not Sketchpad's built-in sine function). Animate P.

TR1D. Add to **TR1C** a similarly constructed graph of a cycle of cosine.

TR1E. In **TR1C**, the drive-point is P on the unit circle, and the domain is restricted to "once around the circle". In order to extend the domain,

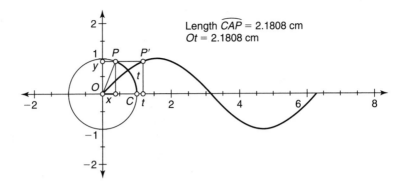

FIGURE 5.2 Graph of $y = \sin t$. The arclength t is laid out from O to t on the horizontal axis. Drag P to vary x and y. Point $P' = (t, y(t))$.

Length $\overset{\frown}{CAP}$ = 2.1808 cm
Ot = 2.1808 cm

sketch a coordinate system, put a moveable point t on the horizontal axis, and apply **Locus** so as to graph the equations $y = \sin t$ and $y = \cos t$.

TR1F. Use **Transform | Translate** and **Measure | Calculate** to confirm the identity $\cos t = \sin(t + \pi/2)$.

SECTION 2 · # The Other Four Basic Trig Functions

Tangent is the name of a function obtained from sine and cosine by the rule $\tan t = \sin t / \cos t$. Since $(\cos t, \sin t)$ is the point (x, y) obtained from an arclength (or angle) of t on the unit circle, it is natural to seek a geometric construction of a segment of length $\tan t$. We already have a triangle with sidelengths x and y, so that the desired length, y/x, can be constructed using similar triangles, as in Figure 5.3.

So far, we have constructions for $\sin t$, $\cos t$, and $\tan t$ as lengths of segments associated with the unit circle. These three together with the remaining three trigonometric functions are listed here. We'll consider the last

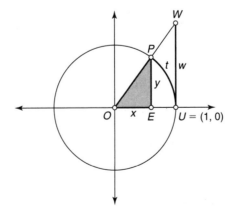

FIGURE 5.3 Triangles *POE* and *WOU* are similar, so that $w/1 = y/x = (\sin t)/(\cos t)$. Conclusion: $|UV| = \tan t$

four in pairs:

$$\sin t = y, \quad \cos t = x, \quad \tan t = y/x, \quad \cot t = x/y, \quad \sec t = 1/x, \quad \csc t = 1/y$$

Tangent and Cotangent

Unlike sine and cosine, the domains of tangent and cotangent are not the set of all real numbers. Since $\tan t = \sin t/\cos t$, values of t for which $\cos t = 0$ are not in the domain of tangent; in particular, there is no $\tan(\pi/2)$. Similarly, the numbers $-\pi$, 0, and π are not in the domain of cotangent.

Considering the quotient $\sin t/\cos t$ in another way, as t varies near $\pi/2$, the denominator is nearly 0 while the numerator is nearly 1. The fraction $\sin t/\cos t$ is accordingly "off the screen". In contrast, $\sin t$ and $\cos t$ are never greater than 1 or less than -1.

ASSIGNMENT 2.1

TR2A. Sketch a segment of length $\tan t$ as in Figure 5.3. Then graph a portion of the tangent function. Don't use the Sketchpad's built-in tangent. Instead follow the example of Figure 5.2, in which a portion of sine appears.

TR2B. Use Sketchpad's built-in tangent function to produce a graph of $y = \tan x$. Sketch the line $y = x$. Select the graph of $y = \tan x$ and verify that Sketchpad does not enable you to reflect the graph in the line. Then place a movable point P on your graph, reflect it to a point P', and sketch the locus of P' as a function of P. Identify the locus of P'.

TR2C. In an xy-coordinate system, use **Measure** to print the slope of the line passing through independent points P and Q. Then use **Measure** to print the tangent of the angle that this line makes with the x-axis.

TR2D. Figure out a construction, using a point (x, y) on the unit circle, that shows a segment of length x/y. This length equals $\cot t$. Emulate **TR2A** to obtain a graph of $y = \cot t$ as t varies from 0 to 2π.

TR2E. Graph the cotangent function using Sketchpad's built-in tangent function. Graph the inverse of the cotangent function using the method of **TR2B**.

Secant and Cosecant

The secant of t, written as $\sec t$, is defined from t units of arclength on the unit circle as $1/x$; that is, $1/\cos t$. A construction of a segment of this length is shown in Figure 5.4. Here, too, as for tangent and cotangent, certain numbers are not in the domains of these functions.

ASSIGNMENT 2.2

TR2F. Emulate Figure 5.4. Add printed measurements which confirm that $|OW| = \sec t$.

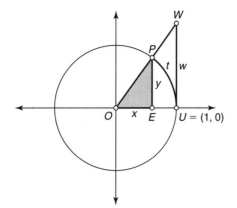

FIGURE 5.4 Triangles *POE* and *WOU* are similar, so that $|OW|/1 = 1/x = 1/(\cos t)$. Conclusion: $|OW| = \sec t$

TR2G. Graph the secant function using Sketchpad's built-in cosine function. Place a movable point *P* on the graph and figure out how to use it with **Translate** and **Locus** to extend your graph to the right.

TR2H. Figure out a construction, using a point (x, y) on the unit circle, that shows a segment of length $1/y$. This equals $\csc t$. Add printed measurements that confirm that the length of your segment agrees with a measure of $1/\sin t$.

TR2I. Graph the cosecant function using Sketchpad's built-in sine function. Extend the result to the right in the manner of **TR2G**.

SECTION 3 # Law of Sines

If a triangle is labeled *ABC* with sidelengths $a = |BC|$, $b = |CA|$, $c = |AB|$, and we write the three vertex angles as *A*, *B*, *C*, then the Law of Sines is the fact that

$$\frac{\sin A}{a} = \frac{\sin B}{b} = \frac{\sin C}{c}$$

An easy proof follows. The altitude from *A* to side *BC* has length $c \sin B$, so the area of $\triangle ABC$, being one-half base times altitude, is $(ac \sin B)/2$. Applying this same area formula using the other two sides as bases gives three expressions for the area:

$$(ac \sin B)/2 = (ba \sin C)/2 = (cb \sin A)/2$$

Dividing by $abc/2$ completes the proof.

ASSIGNMENT 3.1

TR3A. Sketch a triangle *ABC*. Label side *BC* as *a*, side *CA* as *b*, and side *AB* as *c*. (Here, the symbol *a* serves as the name of the segment as well as its length, just as the symbol *A* denotes both a point and an angle.)

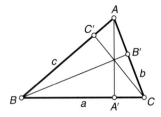

FIGURE 5.5 Triangle ABC with sidelengths a, b, c and altitudes AA', BB', CC'

Keeping in mind that "sin A" abbreviates "sin($\angle CAB$)", use **Measure | Calculate** to print the three quotients that appear in the Law of Sines. Vary $\triangle ABC$ and be sure that the quotients, while varying, stay equal to one another. (See Figure 5.5.)

TR3B. Continuing, what is the constant that those quotients stay equal to? More specifically, what geometric meaning does it have? Add to **TR3A** a printed calculation of $1/(2R)$, where R is the radius of the circumcircle of $\triangle ABC$.

SECTION 4   # Law of Cosines

The Law of Cosines is often written like this: $c^2 = a^2 + b^2 - 2ab\cos C$. It is understood that the symbols refer to a triangle ABC labeled as in Figure 5.5. However, that equation by itself isn't the whole story. Its full meaning becomes clear when we write that the Law also includes

$$a^2 = b^2 + c^2 - 2bc\cos A \quad \text{and} \quad b^2 = c^2 + a^2 - 2ca\cos B$$

To put this another way, it is okay to write the Law as a single equation as long as you remember that it applies to three different "places".

One natural approach to the Law of Cosines starts with the fact that for any given a, b, c that can be sidelengths of a triangle, there is only one triangle that actually has these sidelengths. That is, the vertex angles A, B, C are uniquely determined by a, b, c. Therefore, there must be a way to find the angles in terms of the sides. By the Law of Cosines, for example,

$$\cos A = (b^2 + c^2 - a^2)/2bc, \quad \text{so that} \quad A = \arccos((b^2 + c^2 - a^2)/2bc)$$

If $C = 90°$, then the equation

$$c^2 = a^2 + b^2 - 2ab\cos C$$

reduces to $c^2 = a^2 + b^2$. That is, the Law of Cosines implies the Pythagorean Theorem.

It is not known whether Pythagoras was aware that the converse of this theorem is also true. But let us note that this, too, follows from the Law of Cosines, for in a triangle ABC not already known to be a right triangle, the only way to have both $c^2 = a^2 + b^2 - 2ab\cos C$ and $c^2 = a^2 + b^2$ is if $\cos C = 0$, so that $C = 90°$.

In conclusion, the Law of Cosines includes both the Pythagorean Theorem and its converse, and in addition, the Law tells about all triangles, not just right triangles. Euclid's famous proof of the Law of Cosines is summarized in Figure 5.6.

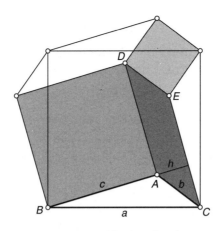

Euclid's proof of the Law of Cosines for obtuse angle A:

$$\begin{aligned} a^2 &= b^2 + c^2 + 2(\text{area of } ACED) \\ &= b^2 + c^2 + 2ch \\ &= b^2 + c^2 + 2c(b\cos(180° - A)) \\ &= b^2 + c^2 + 2bc\cos A. \end{aligned}$$

FIGURE 5.6 Euclid's proof of the Law of Cosines for obtuse angle A

ASSIGNMENT 4.1

TR4A. Start with segments of variable lengths a, b, c. Figure out with pencil and paper some choices of a, b, c for which there is a triangle ABC having a, b, c as sidelengths, and also some choices of a, b, c for which there is no such triangle. For the former case, sketch a triangle having a, b, c as sidelengths. Measure the vertex angle A. Print calculations of $\cos A$ and $(b^2 + c^2 - a^2)/2bc$. Check that they stay equal to each other as you vary a, b, c.

TR4B. Start with a triangle ABC. Print measurements of

$$a^2, \quad b^2, \quad c^2, \quad b^2 + c^2 - 2bc\cos A, \quad c^2 + a^2 - 2ca\cos B, \quad a^2 + b^2 - 2ab\cos C$$

Arrange them in matching pairs on the screen. Vary the shape of $\triangle ABC$ and confirm that matching pairs stay equal.

SECTION 5 Trig Identities

Possibly the single most important trigonometric identity is this:

$$\sin^2 t + \cos^2 t = 1$$

A construction to match this identity has already been given in Section 1, since the birthplace of the functions sine and cosine is the equation $x^2 + y^2 = 1$. The same birthplace provides for *all* other trig identities. In this section, we'll develop sketches that confirm some of the most commonly used identities.

$\sin(s + t) = \sin s \cos t + \cos s \sin t$

Referring to Figure 5.7, suppose angles s and t are placed as shown at the center O of the unit circle, so that

$$s = \angle UOS \qquad t = \angle SOT \qquad s + t = \angle UOT$$

We'll assume for now that s and t are positive and that $s + t < \pi/2$. (After you have a sketch completed, you can vary s and t to confirm that the identity holds for all choices, including negative values.)

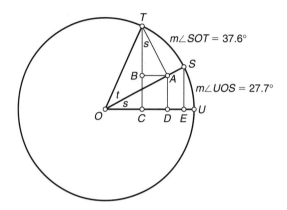

FIGURE 5.7 The identity $\sin(s + t) = \sin s \cos t + \cos s \sin t$

$$\sin(m\angle UOS + m\angle SOT) = 0.908$$
$$\sin(m\angle UOS)\cdot\cos(m\angle SOT) + \cos(m\angle UOS)\cdot\sin(m\angle SOT) = 0.908$$

Continuing with Figure 5.7, by definition of sine and cosine, we have

$$\sin s = |SE| \quad \cos s = |OE| \quad \sin t = |TA| \quad \cos t = |OA|$$

$$\sin(s + t) = |TB| + |BC|$$

In $\triangle BAT$, we have $\cos s = |TB|/|TA|$, so that $|TB| = \sin t \cos s$. Using similar triangles, we also find that $|AD|/|SE| = |OA|/|OS|$, so that $|AD| = \sin s \cos t$. Since $|BC| = |AD|$, we conclude that $\sin(s + t) = \sin s \cos t + \cos s \sin t$.

This is a remarkably productive identity. You should check that each of the following identities can be obtained by a substitution into the identity for $\sin(s + t)$:

- $\sin(s + t) = \sin s \cos t + \cos s \sin t$
- $\sin(s - t) = \sin s \cos t - \cos s \sin t$
- $\cos(s + t) = \cos s \cos t - \sin s \sin t$
- $\cos(s - t) = \cos s \cos t + \sin s \sin t$
- $\sin(2t) = 2 \sin t \cos t$
- $\cos(2t) = \cos^2 t - \sin^2 t$
- $\cos t = \sin(t + \pi/2)$

ASSIGNMENT 5.1

TR5A. Emulate Figure 5.7. Add printed calculations of $\sin(s + t)$ and $\sin s \cos t + \cos s \sin t$, and be sure that these stay equal to each other when you drag points S and T. Then add printed calculations of $\cos(s + t)$ and $\cos s \cos t - \sin s \sin t$.

TR5B. On $\circ(O, |OZ|)$, place movable points P and Q. Rotate Q through $\angle POQ$ about center O, obtaining point Q' satisfying $\angle POQ' = 2 \cdot \angle POQ$. Let $t = \angle POQ$. Print calculations of

$$\sin 2t \qquad 2 \sin t \cos t \qquad \cos 2t \qquad \cos^2 t - \sin^2 t$$

When you vary point Q, ascertain that the double-angle identities are confirmed by the printed calculations.

TR5C. Starting with **TR1A**, color the rectangle that has diagonal from $(0, 0)$ to $(\cos t, \sin t)$. Print a calculation of its area, namely $\sin t \cos t$. Print also $(1/2) \sin 2t$. Animate $(\cos t, \sin t)$, and confirm the identity for $\sin 2t$.

SECTION 6 Waves

Long ago, most users of trigonometry were surveyors, navigators, and civil engineers, for whom the trig functions were most often applied to angles measured in degrees. Today, the main users may be engineers who apply trig functions to numbers (radians) because of the fundamental role of these functions for analyzing wave forms. This is partly because virtually every periodic wave form can be written as a sum of multiples of sines and cosines of multiples of t.

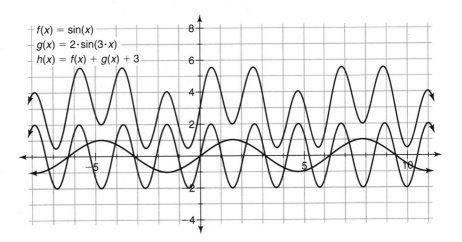

FIGURE 5.8 Waves: graphs of three trig functions

ASSIGNMENT 6.1

TR6A. Graph together the following curves, using Sketchpad's built-in sine and cosine:

$$y = \sin x \qquad y = 2 \sin x \qquad y = 3 \sin x \qquad y = \cos x \qquad y = 2 \cos x$$

Include a caption that explains the role of the coefficient a in the graphs of $y = a \sin x$ and $y = a \cos x$. (Include nonpositive values of a.)

TR6B. Graph together the following, using Sketchpad's built-in sine and cosine:

$$y = \sin x \qquad y = \sin 2x \qquad y = \sin 3x \qquad y = \cos x \qquad y = \cos 2x$$

Include a caption that explains the role of the coefficient b in the graphs of $y = \sin bx$ and $y = \cos bx$.

TR6C. Graph together the following, using Sketchpad's built-in sine and cosine:

$$y = \sin x \qquad y = 2 \sin x \qquad y = 2 \sin 2x \qquad y = \cos x \qquad y = 3 \cos 2x$$

Include a caption that explains the roles of the coefficients a and b in the graphs of $y = a \sin bx$ and $y = a \cos bx$.

TR6D. Using the method for **LO2E** (Figure 3.13, the **parabola grapher**), start with sliders for a, b, c, d, e, f, and sketch a grapher for waves of the form

$$y = a \sin x + b \sin 2x + c \sin 3x + d \cos x + e \cos 2x + f \cos 3x$$

In four captions, tell choices of a, b, c, d, e, f for which your wave "nearly fits" each of the following, for x between 0 and $\pi/2$:

$$y = x$$
$$y = 2 - x^2$$
$$y = 2 \sin 2x - 3 \cos x$$

SECTION 7 Parametric Equations

You are familiar with many curves that are conveniently representable in the form $y = f(x)$. However, this form does not tell where on the curve a moving point is as a function of time. If you want to discuss such a location, or speed or acceleration along a curve, you need a way to represent points on the curve in terms of time; that is, (x, y) must be expressed in terms of a third variable, often denoted by t and called a *parameter*. It turns out that every curve of the form $y = f(x)$ can be just as easily represented in parametric equations, and, on top of that, there are millions of curves that are much more easily represented in this manner than in the form $y = f(x)$.

Simple Curves

The top half of the unit circle, given in the form $y = f(x)$ by $f(x) = \sqrt{1 - x^2}$, is also given by the parametric equations

$$\begin{cases} x(t) = \cos t \\ y(t) = \sin t \end{cases}$$

where t varies from 0 to π. To go around the whole circle, run t from 0 to 2π.

Another simple example is the line passing through two points (x_1, y_1) and (x_2, y_2):

$$\begin{cases} x(t) = x_1 + (x_2 - x_1)t \\ y(t) = y_1 + (y_2 - y_1)t \end{cases}$$

where t varies through the entire number line. In particular, $(x(0), y(0))$ and $(x(1), y(1))$ are the points we started with, and $(x(1/2), y(1/2))$ is their midpoint. If you restrict t to an interval $[a, b]$, then these parametric equations represent the segment from

$$(x(a), y(a)) \quad \text{to} \quad (x(b), y(b))$$

One more advantage of the parametric representation for lines is that vertical lines are included, unlike the case for lines of the form $y = mx$.

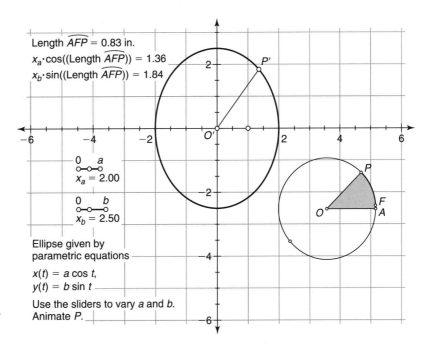

FIGURE 5.9 Ellipse given by parametric equations

ASSIGNMENT 7.1

TR7A. Using the method for **L02E** (Figure 3.13, the **parabola grapher**), start with sliders for a and b, and graph the ellipse

$$\begin{cases} x(t) = a\cos t \\ y(t) = b\sin t \end{cases}$$

Recall that this method of graphing depends on a movable point P that determines t. Animate P.

TR7B. Continuing, use sliders for a and b, and graph together the curves given by

$$\begin{cases} x(t) = a\sec t \\ y(t) = b\tan t \end{cases} \quad \text{and} \quad \begin{cases} x(t) = b\sec t \\ y(t) = a\tan t \end{cases}$$

Animate P.

Lissajous and Super-Lissajous Curves

Lissajous curves are the curves that can be represented by parametric equations of the form

$$\begin{cases} x(t) = a\sin(mt + c) \\ y(t) = b\sin t \end{cases}$$

As t varies from 0 to 2π, the point $(x(t), y(t))$ completes the curve, which must lie inside the rectangle

$$-a \le x \le a \qquad -b \le y \le b$$

since all values of $\sin t$ lie between -1 and 1. The larger m is, the more elaborate the curve.

In order to create artistic Lissajous-like curves, take $x(t)$ and $y(t)$ to be products of several sines and cosines. Figures 5.10 and 5.11 offer such "super-Lissajous curves". Producing these on Sketchpad calls for a technique called *domain-magnification,* since Sketchpad has built-in restrictions on the domains of functions typed into **Graph | New Function**.

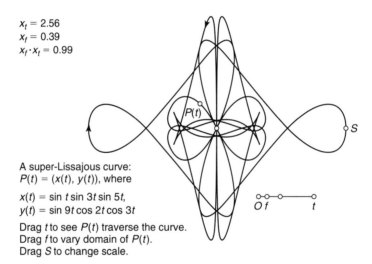

$x_t = 2.56$
$x_f = 0.39$
$x_f \cdot x_t = 0.99$

A super-Lissajous curve:
$P(t) = (x(t), y(t))$, where

$x(t) = \sin t \sin 3t \sin 5t$,
$y(t) = \sin 9t \cos 2t \cos 3t$

Drag t to see $P(t)$ traverse the curve.
Drag f to vary domain of $P(t)$.
Drag S to change scale.

FIGURE 5.10 A super-Lissajous curve

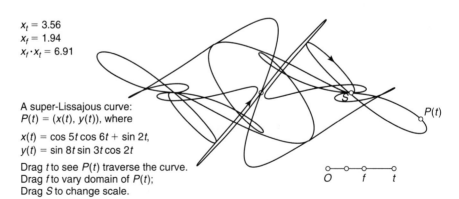

$x_t = 3.56$
$x_f = 1.94$
$x_f \cdot x_t = 6.91$

A super-Lissajous curve:
$P(t) = (x(t), y(t))$, where

$x(t) = \cos 5t \cos 6t + \sin 2t$,
$y(t) = \sin 8t \sin 3t \cos 2t$

Drag t to see $P(t)$ traverse the curve.
Drag f to vary domain of $P(t)$;
Drag S to change scale.

FIGURE 5.11 A super-Lissajous curve

In domain-magnification, we simply multiply Sketchpad's basic variable x by a suitable number, which we'll denote by m. If, for example, x is restricted to the interval $[-10, 10]$, then $10x$ enjoys much greater latitude in the interval $[-100, 100]$. Or, if you have a variable angle (or arclength) t that Sketchpad restricts to "once around the circle"—that's the interval $[0, 2\pi)$—then for $m = 10$, the new variable $10t$ goes around the circle 10 times. This enables us to see "the whole picture" in many cases where, without domain-magnification, only part of the picture would appear.

For those who enjoy generating elaborate periodic symmetric bounded geometric configurations, it is fun to "stitch" a super-Lissajous curve. If t is your parameter (a movable point on a line or circle) and t' is a nearby translate of t, then both $(x(t), y(t))$ and $(x(t'), y(t'))$ traverse the curve. Join these two points by a segment, select t and the segment, and then apply **Construct | Locus**.

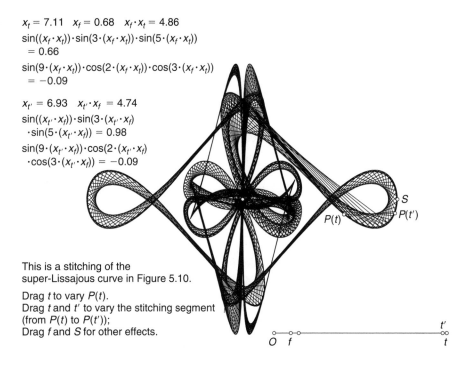

$x_t = 7.11 \quad x_f = 0.68 \quad x_f \cdot x_t = 4.86$

$\sin((x_f \cdot x_t)) \cdot \sin(3 \cdot (x_f \cdot x_t)) \cdot \sin(5 \cdot (x_f \cdot x_t))$
$\quad = 0.66$

$\sin(9 \cdot (x_f \cdot x_t)) \cdot \cos(2 \cdot (x_f \cdot x_t)) \cdot \cos(3 \cdot (x_f \cdot x_t))$
$\quad = -0.09$

$x_{t'} = 6.93 \quad x_{t'} \cdot x_f = 4.74$

$\sin((x_{t'} \cdot x_f)) \cdot \sin(3 \cdot (x_{t'} \cdot x_f)$
$\quad \cdot \sin(5 \cdot (x_{t'} \cdot x_f)) = 0.98$

$\sin(9 \cdot (x_{t'} \cdot x_f)) \cdot \cos(2 \cdot (x_{t'} \cdot x_f)$
$\quad \cdot \cos(3 \cdot (x_{t'} \cdot x_f)) = -0.09$

This is a stitching of the super-Lissajous curve in Figure 5.10.

Drag t to vary $P(t)$.
Drag t and t' to vary the stitching segment (from $P(t)$ to $P(t')$);
Drag f and S for other effects.

FIGURE 5.12 Stitching

Often, when stitching a curve, a second curve will appear. Take a look at Figure 5.13. The outer curve is the one given by the parametric equations in the caption. The inner curve is formed, at least visually, by all those segments. If you could have the segment (or more properly, the line) for every value of t (not just the ones used by Sketchpad as samples), then every point on the inner curve would be tangent to one of the lines. A curve such as this has a name: the *envelope* of the family of lines. Figure 5.13 was drawn using 50 samples.

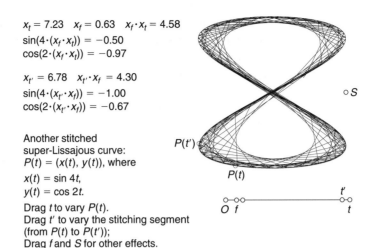

$x_t = 7.23$ $x_f = 0.63$ $x_f \cdot x_t = 4.58$
$\sin(4 \cdot (x_f \cdot x_t)) = -0.50$
$\cos(2 \cdot (x_f \cdot x_t)) = -0.97$

$x_{t'} = 6.78$ $x_{t'} \cdot x_f = 4.30$
$\sin(4 \cdot (x_{t'} \cdot x_f)) = -1.00$
$\cos(2 \cdot (x_{t'} \cdot x_f)) = -0.67$

Another stitched
super-Lissajous curve:
$P(t) = (x(t), y(t))$, where
$x(t) = \sin 4t,$
$y(t) = \cos 2t.$
Drag t to vary $P(t)$.
Drag t' to vary the stitching segment
(from $P(t)$ to $P(t')$);
Drag f and S for other effects.

FIGURE 5.13 Stitching

ASSIGNMENT 7.2

TR7C. Consider Figure 5.10, for which parametric equations are

$$x(t) = \sin t \sin 3t \sin 5t \qquad y(t) = \sin 9t \cos 2t \cos 3t$$

Sketch the curve that is obtained by interchanging $x(t)$ and $y(t)$. But before you do, try to anticipate the result. (You can do it!)

TR7D. Predict and then check how the curve in Figure 5.11 would change if $y(t)$ were changed

from $\sin 8t \sin 3t \cos 2t$ to $-\sin 8t \sin 3t \cos 2t$

TR7E. After examining Figure 5.13, predict the appearance of the curve given by

$$(x(t), y(t)) = (\cos 2t, \sin 4t)$$

Then sketch such a curve, with stitching. Vary the number of samples using **Edit | Advanced Preferences**.

PROJECTS

PROJECT 1: PIECE OF PI

The difference between straight and round may be more mysterious than most people think. Since ancient times, thinkers have studied the fundamental relationship between straightness and roundness. Reduced to its simplest form, the mystery long ago and still today takes this form: *Compare the distance around a circle to the distance across the circle.*

The ancients discovered that the ratio of the distance around to the distance across is between 3 and 4. As time went on, people realized that this ratio is between 3.1 and 3.2, and then later, between 3.14 and 3.15. Modern people, especially some who work with super-computers, have calculated many more digits of this famous ratio. You can see the first fifty million by visiting **http://www.cecm.sfu.ca/projects/ISC/data/pi.html**

If you hadn't previously pondered the mystery of roundness, then ponder this: world-class mathematicians and statisticians are *still* looking for patterns in those millions of digits!

Part 1. Use **Edit | Preferences** to set **Angle Units** to radians and all three **Precision** settings to Hundred Thousandths. Construct several circles of various sizes. Near each circle, print, using **Measure | Calculate**, the distance around the circle (circumference), the distance across the circle (diameter), and the former divided by the latter.

Part 2. Use **Measure | Calculate** to confirm that in a circle of radius r, the arclength subtended by a central angle of measure θ is $r\theta$. Here, and always when an angle is used to formulate length, the angle *must* be measured in "radians"—that is to say, the angle must be a pure number, without units such as centimeters or inches or pixels. What is the connection between **Part 2** and **Part 1**?

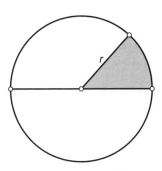

FIGURE 5.14 Piece of Pi

Part 3. Sketch a circle. Let r denote its radius. Now sketch a radius of the circle. (The word radius has two different legitimate meanings, as in the preceding sentences—in one case, radius is a number and in the other, a segment.) Sketch a square, one side of which is the radius you sketched. Predict the ratio of the area of the circle to the area of the square. Use **Measure | Calculate** to check your prediction.

Part 4. Confirm with a sketch and printed measurements that the area of a circular sector of a central angle θ as in Figure 5.14 is $r^2\theta/2$, where r is the radius of the circle.

PROJECT 2: HERON'S FORMULA REVISITED

From Heron's formula (**AN1D**) and the formula

$$area(\triangle ABC) = \frac{1}{2}bc\sin A$$

we can express $\sin A$ in terms of the sidelengths a, b, c of $\triangle ABC$:

$$\sin A = \sqrt{(a+b+c)(-a+b+c)(a-b+c)(a+b-c)}/2bc$$

Confirm this for a variable triangle using **Measure | Calculate**. Then, with pencil and paper, start with this equation for $\sin A$ to derive an equation for $\cos^2 A$ in which the product

$$(a+b+c)(-a+b+c)(a-b+c)(a+b-c)$$

has been completely expanded. Confirm your derivation with printed Sketchpad calculations.

PROJECT 3: CIRCLE, SQUARE, AND ASTROID

The graph of $x^2 + y^2 = a^2$ is a circle. The graph of $x^{2/3} + y^{2/3} = a^{2/3}$ is an astroid. The graph of $|x| + |y| = |a|$ is a square. Can these three curves be seen as examples of a single formulation? In order to answer this question affirmatively, consider the parametric equations

$$x = a\cos^n t, \quad y = a\sin^n t$$

Equivalently,

$$(x/a)^{1/n} = \cos t, \quad (y/a)^{1/n} = \sin t$$

Square both sides of each equation and add the results:

$$(x/a)^{2/n} + (y/a)^{2/n} = \cos^2 t + \sin^2 t = 1$$

so that

$$(x^2)^{1/n} + (y^2)^{1/n} = (a^2)^{1/n}$$

The three cases, circle, astroid, and square, correspond to $n = 1, 3$, and 2, respectively. In order to sketch corresponding graphs, we need to show, simultaneously, graphs of these four sets of parametric equations:

$$x = a|\cos t|^n, \quad y = a|\sin t|^n$$
$$x = a|\cos t|^n, \quad y = -a|\sin t|^n$$
$$x = -a|\cos t|^n, \quad y = a|\sin t|^n$$
$$x = -a|\cos t|^n, \quad y = -a|\sin t|^n$$

Use Sketchpad to plot these separately—or better—plot the first and then obtain the other three as reflections. Once your sketch resembles Figure 5.15, be sure to vary n and note the continuous changes in the appearance of the curve—something dramatic happens when n crosses 0.

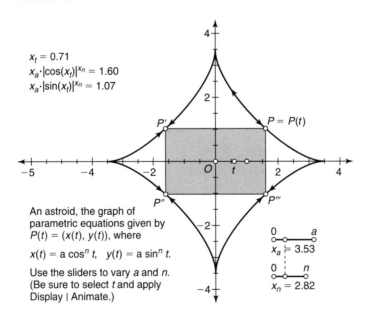

$x_t = 0.71$
$x_a \cdot |\cos(x_t)|^{x_n} = 1.60$
$x_a \cdot |\sin(x_t)|^{x_n} = 1.07$

An astroid, the graph of parametric equations given by $P(t) = (x(t), y(t))$, where

$x(t) = a\cos^n t, \quad y(t) = a\sin^n t.$

Use the sliders to vary a and n. (Be sure to select t and apply Display | Animate.)

$x_a = 3.53$

$x_n = 2.82$

FIGURE 5.15 An astroid

PROJECT 4: TRIG FUNCTION GRAPHERS

Part 1. Using sliders for a, b, c, graph the function $y = a\cos(bx + c)$. Predict the effects of continuously changing each of the variables a, b, c. Then check your predictions by animating each slider. Print your conclusions in captions.

Part 2. Carry out constructions as in **Part 1** using your choice of the following, in place of cos: sin, tan, sec.

Part 3. Start with sliders for each of a, b, c, d, e, f, and construct graphers for the equations

$$y = a\cos(bx + c) \quad \text{and} \quad y = d\cos(ex + f)$$

Drag or animate each of a, b, c, d, e, f and write down observations that seem mathematically notable. Then use your knowledge of trigonometry to refine your statements based on the observations. The ideal is to conclude with clear, short sentences, such as "$\cos(-x) = \cos x$ for all x".

Part 4. Repeat **Part 3** using $y = d\sin(et + f)$ instead of $y = d\cos(et + f)$.

PROJECT 5: THE HYPERBOLIC FUNCTIONS

This project is a preparation for Chapter 9, on Hyperbolic Geometry. The name, *hyperbolic functions,* covers six functions, denoted by

sinh t	csch t
cosh t	sech t
tanh t	coth t

For example, "sinh t" abbreviates "hyperbolic sine of t". You may suspect, from their names, that these functions somehow match the six trig functions. In order to confirm this suspicion, let's review the geometric foundation for the trig functions. Actually, they are often called the *circular functions,* their connection to a circle being clear in Figure 5.1, which shows the circle

$$x^2 + y^2 = 1$$

with the caption, "Birthplace of trigonometry". Likewise, the simple hyperbola with equation

$$x^2 - y^2 = 1$$

serves as a birthplace for the hyperbolic functions. In a manner analogous to the definition of sin t as the y-coordinate of the point (x, y) reached by wrapping t units of arclength around the circle $x^2 + y^2 = 1$, the definition of sinh t is given as follows:

Starting at $(1, 0)$ on the hyperbola $x^2 - y^2 = 1$, wrap counterclockwise t units of arclength; let (x, y) denote the point reached. Then definitions are given by

$$\sinh t = y \quad \text{and} \quad \cosh t = x$$

The remaining four hyperbolic functions are defined by

$$\tanh t = y/x \qquad \operatorname{csch} t = 1/y \qquad \operatorname{sech} t = 1/x \qquad \coth t = x/y$$

Part 1. Graph the hyperbola

$$x^2 - y^2 = 1$$

as a locus of a movable point x on the x-axis, and label the appropriate point as $(\cosh t, \sinh t)$. Here, t denotes arclength on the hyperbola (which could be closely but laboriously approximated as a sum of circular arclengths determined by triples of points placed on the hyperbola).

Part 2. Use **Graph | Plot New Function** to graph of $y = \sinh x$. This function is not included in Sketchpad's menu of functions, but you can use instead the identity

$$\sinh x = (e^x - e^{-x})/2$$

(The "calculus number," e, is offered by Sketchpad's **Graph | New Function | Values.**)

Part 3. In the manner of **Part 2**, graph $y = \cosh x$, using the identity

$$\cosh x = (e^x + e^{-x})/2$$

Part 4. Confirm graphically and prove algebraically in a caption the identity

$$\sinh 2x = 2 \sinh x \cosh x$$

Part 5. Confirm graphically and prove algebraically in a caption the identity

$$\cosh 2x = \sinh^2 x + \cosh^2 x$$

CHAPTER **SIX**

Making Your Own Discoveries

NO DOUBT, YOU'VE wondered "What if?" while carrying out geometric constructions, with or without Sketchpad. If you've done some exploring to answer some of these what-ifs, you may have already made some original discoveries. Such discoveries are thrilling, and there is the possibility that you will, sooner or later, discover something publishable. In this chapter, we'll talk about how discoveries are made, how to cultivate discovery-making and development, how to determine whether a discovery is new enough and good enough to publish, and how to try to get it published.

There are certain mental tendencies exhibited by many innovators, researchers, and other kinds of discoverers. These tendencies often present themselves in the form of themes and variations. Most of this chapter consists of exemplary variations on selected themes. You are urged to seek your own geometric discoveries, largely through variations on themes.

Before commencing with specific themes and variations, a few generalizations are appropriate, pertaining to how one can go about looking for worthy variations. What mental habits serve to spur, guide, and fulfill curiosity?

One such habit is *substitution*. For every property of the circumcircle of a triangle, for example, you can ask if there is an analogous property of the incircle. If something works for an ellipse, then by substitution one asks if something like it works for a hyperbola.

When presented with two or more comparable notions, try *combinations* and *permutations* of the notions. Consider, for example, the notions of isogonal conjugate and inversion in the circumcircle. These are essentially functions, *icon* and *inv,* applicable to a point P. By a combination we mean something like $icon(inv(icon(P)))$, involving two *icons* and one *inv*. By changing the order, as in $icon(icon(inv(P)))$ we have a permutation of the combination. What researchers in every field frequently do is investigate a strategically selected set of permutations of selected combinations.

Another essential habit is *classification*. Given a collection of entities, is there a natural collection of class representatives? For example, among right triangles having integer sidelengths, such as $(3, 4, 5)$, $(6, 8, 10)$, and $(5, 12, 13)$, the first two seem to represent the "same class" (the triangles are similar), whereas the third seems to be of a different class—a class that contains $(15, 36, 39)$, for example. Recognizing classes and good class representatives, such as $(3, 4, 5)$ and $(5, 12, 13)$, is important in the making and developing of discoveries.

Extension from one setting to another is useful. Does something interesting in a 2-dimensional setting extend to a 3-dimensional setting? For example, every triangle has a circumcircle—does every tetrahedron have a circumsphere? Curves in the plane are nicely represented by 2 parametric equations; are curves in 3-space nicely represented by 3 parametric equations? When inverting lines and circles with respect to a given circle, the images are lines and circles—what are the images of ellipses?

Closely related to extension is *generalization*. If some entities have certain forms of representation, ask for a single form that includes the "known" cases along with new ones. For example, the standard forms for conics, such as $(x/a)^2 + (y/b)^2 = 1$, generalize to the form $Ax^2 + Cy^2 + Dx + Ey + F = 0$, which represents all conics with axes parallel to the xy-axes. This form generalizes to $Ax^2 + Bxy + Cy^2 + Dx + Ey + F = 0$, which represents *every* conic in the xy-plane.

The opposite of generalization is *specialization*. This and *exemplification* are especially useful when you must explain your original findings to someone else, or when you are operating in such a thin air of abstraction that you need a "reality check". For example, right triangle, isosceles triangle, and acute triangle are specializations of the notion of triangle. An important kind of exemplification is a *counterexample*. For example, the pedal triangle of the centroid is a counterexample to the (false) proposition that every triangle inscribed in a triangle is perspective to it.

Other habits that accompany discovery-making are *iteration* (e.g., the medial triangle of the medial triangle of the medial triangle), *reversal* (what the discoverer of the anticomplementary triangle did), and *conversing* (asking whether the converse of a proposition is true, as in the case of the Pythagorean theorem).

Ask about *means and extremes*. What is the "average case"? In contrast to the average, what are the extreme cases—such as the greatest possible and the least possible? Of all nonisosceles triangles, is there a "least isosceles" case, so that when you draw it, viewers are less likely to mistake it for a right triangle or equilateral triangle?

An important mental habit in geometry is *dualization,* in which the roles of points and lines are switched. The event "three lines concur in a point" is the dual of the event "three points lie on a line". The duals of the three vertices of a triangle are the three sidelines; for every construction involving the vertices and sidelines, there is a dual construction. For example, when two triangles are perspective, the perspector (center of perspective) is the dual of the perspectrix (axis of perspective).

Another productive habit is *self-referencing.* Consider a thinker who has just learned how to construct the isogonal conjugate of a point P. She applies self-referencing when she asks "What if I apply conjugation to itself; that is, take the isogonal conjugate of the isogonal conjugate?" This question leads to the discovery that $icon(icon(P)) = P$. Another example: how can you "self-reference" the Euler line? A first try, "Euler line of the Euler line", doesn't make sense, but don't give up easily ... what about the Euler lines of three triangles formed by the Euler line? One possibility: let A', B', C' be where the Euler line of $\triangle ABC$ meets sidelines BC, CA, AB, and discover that the Euler lines of triangles $AB'C'$, $A'BC'$, $A'B'C$ form a triangle congruent to $\triangle ABC$ (!).

To summarize, discovery-making depends on curiosity and certain thought-habits, suggested by these verbs: to substitute, combine, permute, extend, generalize, specialize, exemplify, iterate, reverse, dualize, and self-refer. In the following Themes and Variations and in Chapter 7, entitled "Famous Discoveries", you'll find many examples of the results of these thought-habits. To the extent that you already wield these powers, you are already a discoverer. Recognizing these habits and conscientiously applying them to geometric themes can greatly increase your ability to make discoveries and to develop them.

SECTION 1   # Themes and Variations

Imagine a bicycle being pedaled along a line. The front wheel moves forward, and in particular, the center of the wheel moves along a line parallel to the line of motion. Suppose the wheel picks up a tack. As the bicycle continues, the tack will go around, but at the same time, the tack is progressing along with the rest of the bicycle, so its locus is not merely a circle, and we have a discovery kind of question—what sort of locus, or curve, is it?

Theme I: Tack on a Wheel

The theme, "tack on a wheel", serves as a basis for **AN6A** in Chapter 4. The locus constructed there is of a point on a circle rolling on a line. Our objective here is to view this geometric action as a theme that will lend itself to variations:

- the tack—what if it is on the rolling wheel, but not on the rim?
- the line—what if the wheel rolls on a curve, not a straight line?
- the wheel—what if the wheel is elliptical, or some other rollable shape?

Let's take the second of these for further pondering: the simplest curves after lines are circles, so the question takes the form: what if the circles rolls on another circle?

Inasmuch as discovery-making is a kind of search process, one can hardly avoid discerning two kinds of searches, called *depth-first* and *breadth-first*. Think of the theme being searched as a tree, consisting of a trunk, some primary branches from the trunk, secondary branches from the primary ones, tertiaries from secondaries, and so on, out to terminal twigs. In a depth-first search, one follows one branch at a time until reaching a twig, and then backs up just enough to search down to another twig, and so on until finishing the branch, and then repeating this procedure on another primary branch until finishing it, and so on. On the other hand, a breadth-first search examines all the primary branches first, then all the secondary, and so on.

The notion "circle rolls on another circle" is a primary branch. If you ask "what is the locus of the tack if circle 2 rolls on the *outside* of circle 1?" followed by "... if circle 2 is *smaller* than circle 1" then you are starting out on a depth-first search. A breadth-first search starts with "what is the locus if circle 2 rolls on the *outside* of circle 1?" followed by "... if circle 2 rolls on the *inside* of circle 1?" The point here is not that one kind of search is better than the other, but that both search-methods are valuable, and that if you have a penchant for one of them, you should recognize the other and cultivate it, too.

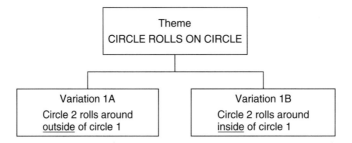

FIGURE 6.1 Beginning a breadth-first search

Suppose that the (fixed) circle 1 has radius r_1 and the (rolling) circle 2 has radius r_2. The circumferences are $2\pi r_1$ and $2\pi r_2$. If the circles have the same size, then they have the same circumference, so that circle 2 completes one revolution exactly upon rolling around circle 1 once. If circle 2 is smaller, it completes more than one revolution about its own center each time it completes a revolution about the center of circle 1. We are led to ask, in terms of r_1 and r_2, how many revolutions, m, circle 2 must roll through in order to complete one revolution about circle 1. Comparing circumferences gives $m(2\pi r_2) = 2\pi r_1$, or simply $m = r_1/r_2$.

The equation $m = r_1/r_2$ shows that if r_1 is an integer multiple of r_2 then when circle 2 rolls just once around circle 1, the "tack" returns to its initial position—thus completing its locus. On the other hand, if, for example, $m = 5/3$, then circle 2 must complete 3 revolutions in order for the locus to be complete. In general, the least number of revolutions of circle 2 needed to complete the locus is the denominator of the fraction obtained by writing r_1/r_2 in lowest terms.

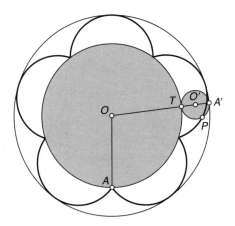

FIGURE 6.2 Locus of tack on circle rolling around outside of a larger circle; ratio 5/1. Drag point T.

In Section 7 of Chapter 5, we gained experience with parametric equations as a way to represent curves for which the representation $y = f(x)$ is inadequate. The circle-rolling-on-circle curves are particularly good examples of curves for which parametric equations are fairly simple and the other form of representation is untenable. By using the parametric equations in sketches, we'll enable a closer search for variations on the underlying theme.

Hypocycloids: circle of smaller radius b rolling **inside** circle of larger radius a:

$$\begin{cases} x(t) = (a - b)\cos t - b\cos((a - b)t/b) \\ y(t) = (a - b)\sin t + b\sin((a - b)t/b) \end{cases}$$

Epicycloids: circle of radius b rolling **outside** circle of radius a:

$$\begin{cases} x(t) = (a + b)\cos t - b\cos((a + b)t/b) \\ y(t) = (a + b)\sin t - b\sin((a + b)t/b) \end{cases}$$

Only the locus of the tack appears in Figure 6.3. In Figure 6.4, the fixed circle and the rolling circle have been included.

The really nice thing about using sliders to control the constants a and b is that we can see "continuous change". A great deal of motion takes place if you keep b fixed and slowly vary a. Can you account for this motion?

ASSIGNMENT 1.1

YO1A. Emulate Figure 6.2 using ratios $r_1/r_2 = 4/1$, $5/1$, $6/1$. (Arrange for changes of ratio to result by dragging A'.) As always, print your observations.

YO1B. Emulate Figure 6.3 using sliders for a and b, and examine the cases $a/b = 3/1$ and $5/3$. Why are the results elaborate when a/b is not a very simple fraction?

YO1C. Add to **YO1B** the rolling inner circle as shown in Figure 6.4.

YO1D. Going back to the method in **YO1A**, can you discover new loci by applying the notion of self-reference to the theme, "tack on a wheel"? If so, compose a sketch that includes such loci, and don't look at the rest of this

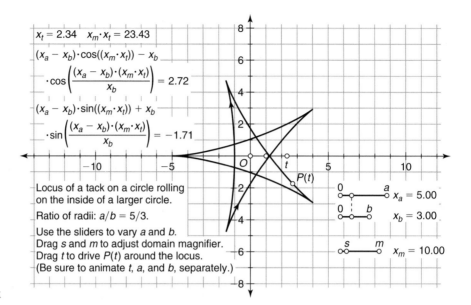

$x_t = 2.34 \quad x_m \cdot x_t = 23.43$

$(x_a - x_b) \cdot \cos((x_m \cdot x_t)) - x_b$

$\cdot \cos\left(\dfrac{(x_a - x_b) \cdot (x_m \cdot x_t)}{x_b}\right) = 2.72$

$(x_a - x_b) \cdot \sin((x_m \cdot x_t)) + x_b$

$\cdot \sin\left(\dfrac{(x_a - x_b) \cdot (x_m \cdot x_t)}{x_b}\right) = -1.71$

Locus of a tack on a circle rolling
on the inside of a larger circle.

Ratio of radii: $a/b = 5/3$.

Use the sliders to vary a and b.
Drag s and m to adjust domain magnifier.
Drag t to drive $P(t)$ around the locus.
(Be sure to animate t, a, and b, separately.)

$x_a = 5.00$

$x_b = 3.00$

$x_m = 10.00$

FIGURE 6.3 Locus of a tack

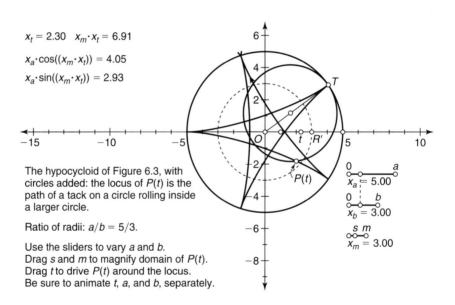

$x_t = 2.30 \quad x_m \cdot x_t = 6.91$

$x_a \cdot \cos((x_m \cdot x_t)) = 4.05$

$x_a \cdot \sin((x_m \cdot x_t)) = 2.93$

The hypocycloid of Figure 6.3, with
circles added: the locus of $P(t)$ is the
path of a tack on a circle rolling inside
a larger circle.

Ratio of radii: $a/b = 5/3$.

Use the sliders to vary a and b.
Drag s and m to magnify domain of $P(t)$.
Drag t to drive $P(t)$ around the locus.
Be sure to animate t, a, and b, separately.

$x_a = 5.00$

$x_b = 3.00$

$x_m = 3.00$

FIGURE 6.4 The hypocycloid in Figure 6.3 with
circles added

paragraph until tomorrow. Otherwise, consider loci associated with a circle
rolling around a circle rolling around a circle. There are many possibilities.
Choose one and compose a sketch that produces such loci.

YO1E. Point P rotates on a circle. While this is going on, line L passing
through point P is rotating about P. A point Q on L, other than point P,
traces a locus. Sketch it. (There are many possibilities. You can pick one,
or better, use a slider to allow a variety of possibilities.)

YO1F. Combine the notion/motion of "tack in a wheel" with the
notion/motion of "inversion in a circle" (Chapter 3, Section 3, and **LO3C**).
(There are many possibilities!)

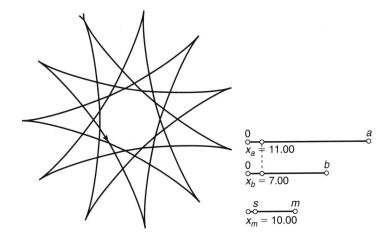

FIGURE 6.5 The hypocycloid having ratio of radii $a/b = 11/7$

$$x_a = 11.00$$
$$x_b = 7.00$$
$$x_m = 10.00$$

$$x_t = -2.42 \quad x_m \cdot x_t = -12.11$$

$$(x_a + x_b) \cdot \cos((x_m \cdot x_t)) - x_b \cdot \cos\left(\frac{(x_a + x_b) \cdot (x_m \cdot x_t)}{x_b}\right) = 6.33$$

$$(x_a + x_b) \cdot \sin((x_m \cdot x_t)) - x_b \cdot \sin\left(\frac{(x_a + x_b) \cdot (x_m \cdot x_t)}{x_b}\right) = 1.07$$

$$x_a = 5.00$$
$$x_b = 2.00$$
$$x_m = 5.00$$

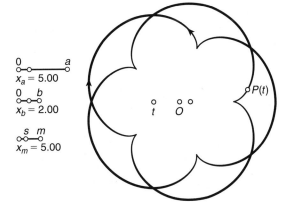

FIGURE 6.6 An epicycloid, the locus of a tack in a circle rolling around the outside of another circle. Ratio of radii: $a/b = 5/2$. Drag t.

Theme 2: Triangle Centers

A circle has only one center. A square has only one center. A rectangle has only one center. But triangles are different; they have many centers. In earlier chapters, you encountered special points whose names are based on the notion of centrality. These include *cent*roid, circum*center*, ortho*center*, in*center*.

Surprisingly, it was only during the past few years that a general definition of triangle center was published. That definition is technical and need not concern us here. For present purposes, you can call a point P in the plane of $\triangle ABC$ a triangle center if there is a construction of P such that, if the roles played by the vertices A, B, C are rearranged in any order, the resulting point remains unchanged. Here, the triangle ABC is understood to be an arbitrary triangle, so that P can be regarded as a function of the variables *a*, *b*, *c* (the sidelengths). Since sin *A*, sin *B*, sin *C* are respectively proportional to *a*, *b*, *c* (by the Law of Sines), we may just as well regard *P* as a function of the angles *A*, *B*, *C*.

With regard to constructions of triangle centers, we shall employ the term *cyclically defined*—if you've constructed an object *U* following instructions using the symbols *A*, *B*, *C*, then, when you repeat the construction with the same instructions but with *A*, *B*, *C* replaced by *B*, *C*, *A* respectively, the new object is cyclically defined from *U*. If you represent the original object with the functional representation *U*(*A*, *B*, *C*), then the new object is *U*(*B*, *C*, *A*). Almost always, we'll deal with yet another new object as well, and that is *U*(*C*, *A*, *B*), cyclically defined from *U*(*B*, *C*, *A*). If you apply the cyclic transformation to *U*(*C*, *A*, *B*), you get *U*(*A*, *B*, *C*) which is the original object. An example is *U*(*A*, *B*, *C*) = median through *A*, from which follows

> *U*(*B*, *C*, *A*) = median through *B*
> *U*(*C*, *A*, *B*) = median through *C*

In Chapter 1, we witnessed 3 medians concurring in the centroid, 3 altitudes concurring in the orthocenter, and 3 angle bisectors concurring in the incenter. In all these cases, the lines are of the form *AA'*, *BB'*, *CC'* where *A'B'C'* is a "nice" triangle, and the corresponding triangle center is the perspector of the two triangles, *ABC* and *A'B'C'*. Accordingly, our search-method, and the theme of this section, is the notion of nice, or *successful*, triangles—any triangle *A'B'C'* for which the lines *AA'*, *BB'*, *CC'* concur. Figure 6.7 offers one of millions of possible successful triangles *A'B'C'*.

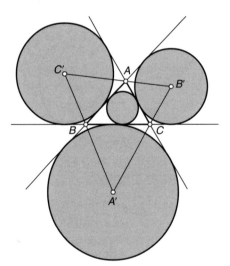

FIGURE 6.7 Triangle *ABC* and the triangle *A'B'C'* formed by the excenters

ASSIGNMENT 1.2 **YO1G.** Emulate Figure 6.7 in a sketch. Determine experimentally whether △*A'B'C'* appears to be successful. Then consider the points in which the inner circle touches the triangle: do they offer a successful triangle? Seek another candidate in Figure 6.7 and check it.

YO1H. Let I be the incenter of $\triangle ABC$. Let $A' = AI \cap BC$, and define B' and C' cyclically. Let $A'' = B'C' \cap BC$, and define B'' and C'' cyclically. Let U_A be the circle with diameter $A'A''$, and define U_B and U_C cyclically. Confirm that the three circles concur in two points. Regard this concurrence as a theme, and seek notable variations.

YO1I. Don't read this paragraph until you've worked on **YO1G** for a while. Let A' be the point of tangency of the incircle with line BC, and let B' and C' be cyclically defined. Confirm that the triangle $A'B'C'$ is successful. Let A'' be the point of tangency of the A-excircle with line BC, and let B'' and C'' be cyclically defined. Confirm that the triangle $A''B''C''$ is successful. The perspector in the first case is called the *Gergonne point;* in the second, the *Nagel point.*

YO1J. Continuing from **YO1G**, let A'' be the point where the perpendicular to AB through the B-excenter meets the perpendicular to AC through the C-excenter. Define B'' and C'' cyclically. Confirm that $\triangle A''B''C''$ is successful.

YO1K. Continuing from **YO1G**, let A_B be the point where the B-excircle meets line AB, and A_C the point where the C-excircle meets line AC. Define B_C, C_A cyclically from A_B, and define B_A, C_B cyclically from A_C. Let A' be the midpoint of segment $A_B A_C$, and define B' and C' cyclically. Is $\triangle A'B'C'$ successful?

YO1L. Continuing **YO1K**, let $A' = A_B C_A \cap B_A A_C$, and define B' and C' cyclically. Is $\triangle A'B'C'$ successful?

YO1M. Don't read this paragraph until you've worked on **YO1H** for a while. The triangle centers constructed in **YO1H** are called the *isodynamic points.* If, at the beginning of **YO1H**, you substitute an independent point P for the incenter I, you'll again find three circles concurrent in two points for many positions of P. Confirm this, and also discover whether the centers of the three circles are collinear. If so, seek notable properties of their line.

Theme 3: The Line at Infinity

No contradictions result if a "line at infinity" is attached to the plane of a given triangle. In fact, quite the opposite is true: many interesting objects and relationships can be accounted for in terms of this line. If you have two parallel lines, they meet in a point on the line at infinity. In fact, all the lines that have the same direction meet in a single point on the line at infinity. For this reason, it is helpful to think of each point on this special line as a direction in the plane.

Recall (from Chapter 3, Section 4) that if P is a point in the plane of $\triangle ABC$, and P doesn't lie on a sideline, then its isogonal conjugate is constructed as follows: reflect line AP about the angle bisector at A; reflect line BP about the angle bisector at B; reflect line CP about the angle bisector at C; then the three reflected lines concur in the desired isogonal conjugate of P. For convenience, we'll write the isogonal conjugate of P as P^{-1}. In Chapter 3, we've already seen that the isogonal conjugate of a point P on the circumcircle of $\triangle ABC$ is on the line at infinity. What happens when you construct the three reflected P-lines is that they are parallel—hence they meet in a point on the line at infinity. Although you can't see the point, you can see the direction indicated by the three parallel lines.

Figures 6.8–6.10 indicate lines through points P on the circumcircle, each aimed in the direction of P^{-1}.

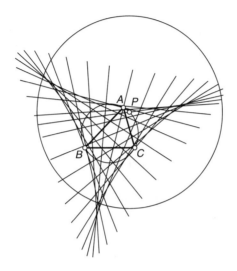

FIGURE 6.8 Locus of lines (actually segments) in the direction from P to the isogonal conjugate of P. Here, Sketchpad used 20 sample positions for P.

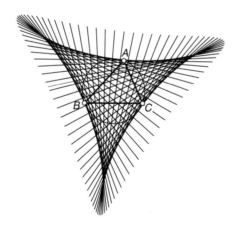

FIGURE 6.9 Locus of lines (segments) in the direction from P to the isogonal conjugate of P. Here, Sketchpad used 50 sample positions for P.

The number of points P sampled can be changed using **Edit | Advanced Preferences**. To access these options, click **Edit** and press the **Shift** key; at the bottom of the **Edit** menu, you'll see **Preferences** replaced by **Advanced Preferences**.

In the **Advanced Preferences** dialog box, click on **Sampling**. You'll see **Number of Samples** for **New Point Loci**, for which the maximum is 10000. You should try out a number of different choices. To see a new number of samples, you must delete the previous locus, and then, after typing in the new number, apply **Locus** again. (What would you expect to see if you requested 10000 samples of P for the locus in Figure 6.8?)

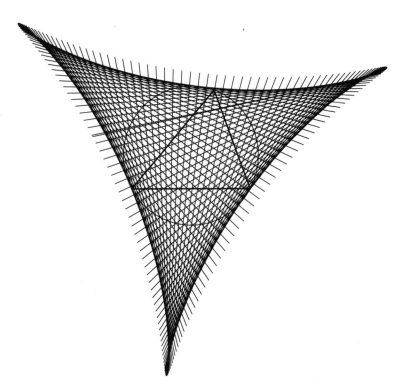

FIGURE 6.10 Locus of lines (segments) in the direction from P to the isogonal conjugate of P. Here, Sketchpad used 100 sample positions for P.

ASSIGNMENT 1.3

YO1N. Start with $\triangle ABC$ and its circumcircle. Place a movable point P on the circle, and carry out the construction of the three parallel lines AP^{-1}, BP^{-1}, and CP^{-1}.

YO1O. Add to **YO1N** an independent line L. Construct line L_A through point A parallel to L, line L_B through point B parallel to L, and L_C through point C parallel to L. Reflect L_A about the angle bisector of A, and likewise for L_B and L_C. Let Q be the point where the three reflected lines concur. Drag L around the screen and confirm that its child, Q, stays on the circumcircle of $\triangle ABC$. What does this illustrate about isogonal conjugates?

YO1P. Suppose L is a line and Q is its child as constructed in **YO1O**. Suppose L' is another such line, with child Q'. Discover as many pairs of lines L and L' as you can for which the points Q and Q' are opposite ends of a diameter of the circumcircle of $\triangle ABC$. Determine how L and L' are related to each other.

Theme 4: Equilateral Triangles

Certain equilateral triangles associated with a variable triangle ABC are very interesting. In this section, we'll construct several of the known equilateral triangles, with a view toward discovering relationships among them and perhaps new equilateral triangles.

Probably the most famous such equilateral triangle is the Morley triangle, introduced in Chapter 1, Section 3. Closely associated with the Morley triangle, which we'll now call the *1st Morley triangle,* are two others.

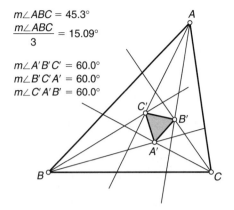

$m\angle ABC = 45.3°$

$\dfrac{m\angle ABC}{3} = 15.09°$

$m\angle A'B'C' = 60.0°$
$m\angle B'C'A' = 60.0°$
$m\angle C'A'B' = 60.0°$

FIGURE 6.11 Triangle ABC and its 1st Morley triangle

ASSIGNMENT 1.4

YO1Q. First, construct the 1st Morley triangle, as in Figure 6.11. A 2nd Morley triangle has for A-vertex the point of intersection of the first trisector of angle $(180° + \angle B)$ and the first trisector of angle $(180° + \angle C)$. (You could, for example, obtain the first of these trisectors by rotating B about A using angle measurement $60° + (\angle ABC)/3$.) The other two vertices of the 2nd Morley triangle are defined cyclically. Construct the 2nd Morley triangle, and print measurements that confirm that it is equilateral.

YO1R. Add to **YO1Q** a 3rd Morley triangle, using trisectors of angles

$$360° + \angle A, \quad 360° + \angle B, \quad 360° + \angle C$$

Then seek noteworthy relations among the three Morley triangles and $\triangle ABC$.

YO1S. Figure 6.10 suggests that as P goes around the circumcircle, there are three locations for which the line PP^{-1} is tangent to the circumcircle. These three points are the vertices of the *circumtangential triangle*. Its sides are pairwise parallel to the sides of the Morley triangle. Use this fact and **YO1Q** to construct the circumtangential triangle. Print measurements that confirm that the circumtangential triangle is equilateral. Seek other equilateral triangles in connection with Figure 6.10.

YO1T. Seek points P whose pedal triangle or antipedal triangle (Chapter 2, Section 3) is equilateral.

SECTION 2 # The Publishing of Discoveries

Every year, thousands of new mathematical discoveries are published. Hundreds of them are geometric, and dozens are within the realm of Euclidean geometry. A substantial portion of them come from students, as well as engineers, chemists, and physicians.

In this section, we'll talk about how one should go about trying to get published. Let's assume that you've found something that can be stated in less than four paragraphs and thoroughly explained in less than twenty paragraphs. The first three requirements for publication is that the discovery be *interesting, new* to the readers of the intended publication, and *well written*.

Concerning the criteria—interesting, new, and well written—the best way to gain a sense of what these adjectives mean is to familiarize yourself with published material that exemplifies the kind of discovery that you might be considering submitting. The greatest number of modern discoveries in Euclidean geometry are published as main articles, notes, or, especially, as problem proposals in journals such as these:

American Mathematical Monthly
http://www.maa.org/pubs/monthly.html

College Mathematics Journal
http://www.maa.org/pubs/cmj.html

Crux Mathematicorum
http://journals.cms.math.ca/CRUX/

Forum Geometricorum
http://forumgeom.fau.edu/

Mathematical Gazette
http://www.m-a.org.uk/eb/mg/index.htm

Mathematics Magazine
http://www.maa.org/pubs/mathmag.html

Proposers and solvers of problems that appear in these journals include many university students. Here's how it works:

1. Readers contribute new problems in the form of proposals
2. An editor receives attempted solutions from readers
3. One or more solutions are selected and published

This method of publication reaches many people and establishes priority, or intellectual ownership, of your discovery, in case it is new to the literature.

Discoveries for which there is a classroom slant often find their way into articles and letters to the editor of *The Mathematics Teacher* and other journals for teachers of mathematics.

There are many websites that specialize in online problem posing and solving. You can find such sites quickly through Internet searching. Internet-published discoveries, at the time of this writing, are generally not subject to as much refereeing and not as securely "yours" as traditionally published discoveries.

On the other hand, during the past few years, several new fully refereed journals that carry full-length articles have appeared. One of these that specializes in Euclidean geometry is *Forum Geometricorum,* listed above.

Many problem-collections rich in Euclidean geometry, especially in connection with national and international problem-solving competitions, are accessible from **http://camel.math.ca/Exams/competitions.html**.

These competitions are for problem solving rather than the making of discoveries; however, the thousands of problems archived in these sources provide gold mines of "themes" having potentially new variations, as well as examples of those aforementioned prerequisites for publishing: interesting, new, and well-written.

Among books, an excellent resource is Stanley Rabinowitz's *Index to Mathematical Problems 1980–1984*, MathPro Press, 1992. If your library doesn't own a copy, it would be worthwhile to borrow one through an interlibrary loan. New and interesting geometry problems published around the world during 1980 to 1984 are restated on pages 122–184 and are cross-referenced in various ways.

Following are four problems quoted in Rabinowitz's *Index.* By consulting the *Index,* you can locate solutions. However, our interest here is not primarily in solutions. Rather, the problems are chosen for two reasons: to exemplify interesting, new, well-written discoveries, and for the experience of designing and experimenting with sketches that illustrate the problems in action.

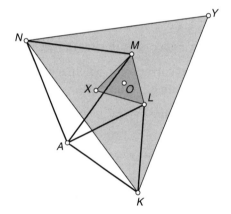

FIGURE 6.12 Rotate or change size of equilateral triangles *AKL* and *AMN* by moving *K* and *N*. Equilateral triangles *LMX* and *KNY* remain concentric.

ASSIGNMENT 2.1

For each of these problems, create a sketch that confirms the discovery stated in the problem.

YO2A. What is the locus of the third vertex of all equilateral triangles having one vertex on a given circle and a second vertex at a fixed point outside the circle?*

YO2B. Let *ABC* be an equilateral triangle with center *O*. Prove that if *P* is a variable point on a fixed circle with center *O*, then the triangle whose sides have lengths |*PA*|, |*PB*|, |*PC*| has a constant area.**

---

*Walt Ainsworth, Problem 3858, *School Science and Mathematics* 81 (1981) 438.

**Stanley Rabinowitz, Problem 728, *Crux Mathematicorum* 8 (1982) 210.

YO2C. Let *AKL* and *AMN* be equilateral triangles. Prove that the equilateral triangles *LMX* and *KNY* are concentric (if *Y* is on the properly chosen side of *NK*). See Figure 6.12.*

YO2D. Let the triangle *ABC* be inscribed in a circle and let point *P* be a point in the interior of the circle. The line segments *AP*, *BP*, and *CP* are extended to meet the circle in points *D*, *E*, and *F*, respectively. Describe all such *P* for which**

$$\frac{|AP|}{|PD|} + \frac{|BP|}{|PE|} + \frac{|CP|}{|PF|} \leq 3$$

PROJECTS

- -

PROJECT 1: THEMES FOR VARIATIONS

"In two minutes, tell as many possible uses of a brick as you can". This well-known test for creativity is much like a certain type of mathematical discovery-mindedness, except that in pure mathematics, the notion of "possible uses" is liberated from physical restrictions. Instead of "possible uses", a mathematician seeks "interesting connections" within mathematics, preferably connections that are simple but surprising, or, to use the words popular with many mathematicians: *natural* but *unexpected*. In Project 1, the theme (or brick) is a point *P* going around a circle, and you are asked to tell as many possibly interesting loci generated from *P* as you can. **Parts 1–3** offer examples that can serve for further original exploration, and **Part 4** is more open-ended.

Part 1. Sketch a circle *c*, centered at the origin of the *xy*-plane, and place a movable point *P* on *c*. Figure out how to sketch a second circle, *c'*, tangent to *c* at *P* and also tangent to the *x*-axis. Then sketch and discuss the locus of the center of *c'*.

Part 2. Sketch a circle with center on the line *y* = *x* in the *xy*-plane, and place a movable point *P* on *c*. Sketch and discuss the locus of the midpoint between the *x*- and *y*-intercepts of the line tangent to the circle at *P*.

Part 3. Sketch a circle *c* centered at the origin of the *xy*-plane, and place a movable point *P* on *c*. Let *A* be the projection of *P* onto the *x*-axis, and *B* that on the *y*-axis. Let *f*(*P*) be the projection of *P* onto segment *AB*. Sketch and discuss the locus of *f*(*P*).

Part 4. On a clean sheet of paper, develop plans for an original locus to be generated by a rotating point *P*. Don't try to foresee how the locus will appear, don't be distracted by any notion of usefulness, and don't worry about whether the locus will be "known". After making rough sketches, develop plans for another locus. And another. Then select one of the

---

*Jordi Dou, Problem E2866, *American Mathematical Monthly* 88 (1981) 66.
**Peter Ørno, Problem 1120, *Mathematics Magazine* 55 (1982) 182.

ideas to be developed into a sketch. When the sketch is finished—or before—look for variations and other possibilities. Check them out. Feel free to introduce additional independent lines, circles, bisectors, and so on. Develop at least three sketches of original loci.

PROJECT 2: CONSTANT AREA OR PERIMETER

Part 1. In an *xy*-coordinate system, adjust the scale so that the unit point is 1 inch from the origin *O*. Sketch *B* on the positive *x*-axis and *H* on the positive *y*-axis. Place a movable point *b* on the *x*-axis, and let *h* be the point on the *y*-axis such that area($\triangle bOh$) = area($\triangle BOH$). Sketch the midpoint *m* of segment *bh*. Sketch and discuss the locus of *m* as a function of *b*.

Part 2. Let *P* be a point to be rotated on a circle with center *O*. Sketch a line *L* that is tangent to the circle and is also parallel to line *OP*. Complete the parallelogram having segment *OP* as a side and another side lying within line *L*. Let *R* be the vertex of the parallelogram that is opposite vertex *O*. Sketch the locus of point *R*.

Part 3. In an *xy*-coordinate system, adjust the scale so that the unit point is 1 inch from the origin *O*. Sketch point *U* on the negative *x*-axis and point *V* on the negative *y*-axis. Let *P* be a movable point on the positive *x*-axis. In the vertical line through *P*, sketch the point *Q* satisfying

$$\text{perimeter}(POQ) = \text{perimeter}(UOV)$$

Sketch the locus of point *Q*.

Part 4. Sketch a triangle *ABC*. Sketch the locus of a point *P* such that the triangles *PBA* and *PAC* have equal perimeters.

PROJECT 3: COMBINED ROTATIONS

In **Y01E**, the locus of a point rotating about a rotating point has been sampled. In Project 3, we wish to extend this combination of rotations yet another step—with results that are not only good for discovery-making, but also fun to work with.

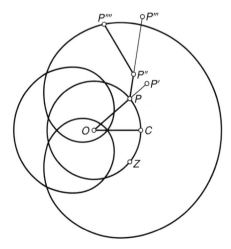

FIGURE 6.13 Combined rotations

Part 1. On $o(O, |OZ|)$, sketch a point C and a movable point P. Sketch a movable point P' on segment OP so that P is between O and P'. Let P'' be the rotation of P' about P using $\angle COP$. Sketch a movable point P''' on segment PP'' so that P''' is between P and P''. Let P'''' be the rotation of P''' about P'' using $\angle P'PP''$. Sketch the locus of P''''.

Part 2. Modify your sketch for **Part 1** by changing the angles of rotation. Be sure to animate not only P, but also the points P' and P'''. The objective is to produce loci that seem attractive for reasons of symmetry and for subjective reasons. The mathematical value in this case is not so much in the properties of the loci, but rather, what you have to do with Sketchpad in order to produce whatever effects you wish to produce. In your modifying and experimenting, feel free to use translations, reflections, and dilations, in addition to rotations.

PROJECT 4: COMBINED TRANSLATIONS

Part 1. Inside $\triangle ABC$, sketch $o(O, |OZ|)$ and a movable point P on the circle. Sketch an independent point A_1 away from $\triangle ABC$. Let B_1 be the translation of P by vector AP, let C_1 be the translation of B_1 by vector BP, and let P_1 be the translation of C_1 by vector CP. Thus, point P_1 is the translation of P by the vector $AP + BP + CP$. Sketch the locus of P' as a function of P. Next, sketch the perpendicular projections D, E, F of P onto sidelines BC, CA, AB. Sketch an independent point A_2 away from your other objects, and let P_2 be the translation of P by the vector $PD + PE + PF$. Sketch the locus of P_2.

Part 2. The locus of P_2 in **Part 1** is an ellipse. Figure out a vector sum (i.e., combined translations) involving O that carries point A_2 onto the center of the ellipse.

Part 3. In **Part 1**, free P from $o(O, |OZ|)$ using **Edit | Split**, and seek a location of point P such that $AP + BP + CP$ is the zero vector. Then seek a location for point P such that $PD + PE + PF$ is the zero vector. (The two locations are well-known triangle centers; determine which ones.)

Part 4. Use **Parts 1–3** as theme-material for original variations involving combined translations.

PROJECT 5: THE PHILO LINE

The search for your own discoveries can be especially rewarding if the theme is a topic that hasn't been exhausted by previous discoverers. Such a theme is the Philo (or Philon) line. If a point P' is placed in the interior of an angle having edges d and e, then the shortest segment from d through P' to e is the *Philo line* of P' for the given angle. Among the few books that note that this segment is not Euclidean constructible is Howard Eves's *A Survey of Geometry*, vol. 2, Allyn and Bacon, 1963, pages 39 and 234–236.

    Part 1 starts with an angle and a segment passing across it, and gives a Euclidean construction for the point P' whose Philo line is the given segment. In **Part 2** the minimality of the length of the segment is confirmed. In **Part 3** the locus of P' is confirmed to have a certain parametric representation, and in **Part 4** a related locus is sketched. The first four Parts demonstrate that two interesting curves, possibly not previously published, arise in connection with the ancient but now again fresh topic of Philo line. **Part 5** is an invitation to seek other interesting curves and properties associated with Philo lines.

Part 1. Let A be a movable point on the positive x-axis in an xy-coordinate system having origin O. Let B be an independent point in Quadrant I. Let Z be a movable point on the negative x-axis, and let P be a movable point on $o(O, |OZ|)$. (Until the sketch is finished, keep P inside $\angle AOB$.) Let a and b be the points in which the line tangent to $o(O, |OZ|)$ meets the lines OA and OB, respectively. Let P' be the translation of a by vector Pb. Sketch the locus of P' as a function of P.

Part 2. Confirm that when P lies within $\angle AOB$, of all segments that pass through P' and have endpoints on segments OA and OB, segment ab is the shortest.

Part 3. Confirm using **Graph | Plot As (x,y)** that the locus of P' is given parametrically by the equations

$$x = r(\sec\theta + \sin\theta\tan(\theta - \theta_0)) \quad y = -r\cos\theta\tan(\theta - \theta_0)$$

where

$$r = |OZ| \quad \theta = \angle AOP \quad \theta_0 = \angle AOB$$

Part 4. In your sketch for **Part 1**, insert (1) the perpendicular to line OB through b, (2) the perpendicular to line OA through a, and (3) the perpendicular to line ab through P'. These three perpendiculars concur in a point. Label it Q and sketch its locus as a function of P. Confirm using **Graph | Plot As (x,y)** that the locus of Q is given parametrically by the equations

$$x = r\sec\theta \quad y = -r\sec\theta\tan(\theta - \theta_0)$$

Part 5. Seek original discoveries involving one or more Philo lines. Consider, for example, the fact illustrated by **Y01G–Y01M** that an arbitrary triangle has many special points, but that, possibly, no one has discovered a special point whose definition depends on the three Philo lines of a triangle.

PROJECT 6: PARALLEL LINES

Here is an interesting proposition: given any three parallel lines, there exists an equilateral triangle such that each of the lines passes through one vertex of the triangle, as typified by Figure 6.14.

Part 1. Prove the proposition, or else find a Sketchpad method that can convince your friends that the proposition is true.

Part 2. Write down several variations of the proposition. After experimenting and refining your statements, list them clearly in an order that lends itself to explaining them to a group of listeners. (Combinations of the following are open for variation: *three, parallel, lines, equilateral triangle*.) Illustrate two of your variations with Sketchpad constructions.

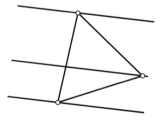

FIGURE 6.14 A vertex on each line

CHAPTER SEVEN

Famous Discoveries

FAMOUS DISCOVERIES THAT invite further exploration include the theorems of Pythagoras, Ptolemy, Pascal, Poncelet, and others represented in this chapter. We begin with one of the world's most famous geometric lines.

SECTION 1

Distance Ratios Along the Euler Line

Recall from Chapter 1 that the Euler line is determined by any two of several triangle centers that lie on it, including the centroid (G), circumcenter (O), and orthocenter (H). In **SS1K**, attention has already been called to the constancy of the ratio $|GH|/|GO|$ as the shape of the reference triangle ABC varies. This constancy has been known since ancient times.

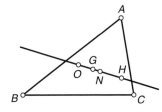

FIGURE 7.1 Euler line with labeled points G, O, H, N

The Points G, O, H, N

The nine-point center (N in Figure 7.1), defined as the center of the nine-point circle, is easily sketched as the circumcenter of the medial triangle. It, too, lies on the Euler line and "fits in" with G, O, and H.

ASSIGNMENT 1.1

FD1A. Start with a triangle ABC. Use tools from Chapter 2, or else start from scratch, to sketch the points G, O, H, and N. Sketch the Euler line. Ascertain that G, O, H, N remain collinear when A, B, or C is dragged. As always, print your observations.

FD1B. Continuing from **FD1A**, use **Measure | Calculate** to print these ratios:

$$|GH|/|GO| \qquad |GN|/|GO| \qquad |GH|/|GN| \qquad |OH|/|ON|$$

Let H' be the reflection of H in O; predict and confirm the value of the ratio $|GH'|/|GH|$. (The point H' is the *de Longchamps point*. If you care to research this further, use **Google.com** to search for any triangle center that has a well established name, such as this one, and you will get several hits.)

More Points on the Euler Line

Recall that the inverse of a point in a circle is collinear with the point and the center of the circle. Accordingly, if P is on the Euler line, then its inverse in the circumcircle is also on the Euler line.

ASSIGNMENT 1.2

FD1C. Add to **FD1B** the inverses of G and H in the circumcircle. Label these inverses as G' and H'. Determine the points X among G, O, H, N for which the distance-ratio $|XH'|/|XG'|$ stays constant as the triangle ABC varies. Starting with such a point X, sketch and confirm a point Y, other than H', for which $|XY|/|XG'|$ stays constant. Confirm this constancy.

FD1D. Use **FD1C** to determine if the ordering of the six points

$$G, \ O, \ H, \ N, \ G', \ H'$$

stays the same for all triangles ABC. What "strings" (such as H, N, G, O) from this collection of points appear to stay in order for all shapes of the variable triangle ABC?

SECTION 2

Feuerbach's Theorem, Point, and Triangle

Among the most famous geometric discoveries of the early nineteenth century is *Feuerbach's theorem: the incircle and the nine-point circle* (Figure 2.3) *meet in exactly one point, and also, each of the three excircles meets the nine-point circle in exactly one point.* The first of these four points is known as the *Feuerbach point,* and the other three are the vertices of the *Feuerbach triangle.*

ASSIGNMENT 2.1

FD2A. Start with a triangle ABC. Sketch the Feuerbach point and the Feuerbach triangle. (If Sketchpad falters on the point of tangency of two circles, use the fact that the Feuerbach point is the point in which the line of the incenter and the nine-point center meets the incircle.)

FD2B. Continuing, hide the excenters and excircles. Then confirm that the Feuerbach triangle is perspective to $\triangle ABC$.

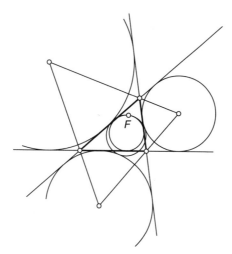

FIGURE 7.2 Feuerbach configuration: the nine-point circle is tangent to the incircle and the three excircles

FD2C. Starting with **FD2A**, hide the excenters and excircles. Then confirm that the Feuerbach triangle is perspective to the cevian triangle of the incenter.

SECTION 3 # Ptolemy's Theorem

Three points determine a circle. Therefore, if you start with four points, maybe they all lie on a circle, and maybe they don't. Ptolemy of Alexandria pondered this observation and discovered a theorem about it. The theorem is that *a quadrilateral ABCD (as in Figure 7.3) can be inscribed in a circle if and only if the sides and diagonals satisfy this equation:*

$$|AB| \cdot |CD| + |AD| \cdot |BC| = |AC| \cdot |BD|$$

$AB = 2.60$ in. $AB \cdot CD + AD \cdot BC = 7.92$ in^2
$CD = 2.00$ in. $\qquad AC \cdot BD = 7.92$ in^2
$AD = 2.34$ in.
$BC = 1.17$ in.
$AC = 2.94$ in.
$BD = 2.70$ in.

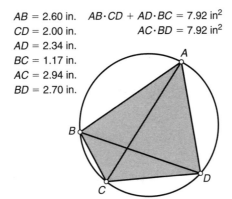

FIGURE 7.3 Ptolemy configuration: quadrilateral inscribed in a circle

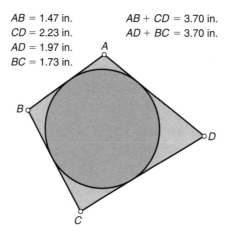

AB = 1.47 in.   AB + CD = 3.70 in.
CD = 2.23 in.   AD + BC = 3.70 in.
AD = 1.97 in.
BC = 1.73 in.

FIGURE 7.4 Quadrilateral circumscribing a circle

Such a quadrilateral is called a *cyclic quadrilateral*. Substituting into Ptolemy's theorem the notion of "circumscribed" for "inscribed", we obtain another ancient theorem: *a quadrilateral ABCD (as in Figure 7.4) can be circumscribed about a circle if and only if the sides satisfy this equation:*

$$|AB| + |CD| = |AD| + |BC|$$

FD3A. Sketch movable points labeled A, B, C, D in counterclockwise order on a circle. Use **Measure | Calculate** to print measurements that confirm Ptolemy's theorem. Predict what will happen when you animate A. What does happen?

FD3B. Continuing from **FD3A**, sketch the midpoints of the four sides of quadrilateral $ABCD$. Through each midpoint, sketch the line perpendicular to the opposite side of the quadrilateral. The four perpendiculars concur in a point. Label it P and sketch its locus as a function of A.

FD3C. Start with $\circ(O, |OZ|)$. Sketch a quadrilateral that circumscribes your circle. Label its vertices A, B, C, D in counterclockwise order. Use **Measure | Calculate** to confirm that

$$|AB| + |CD| = |AD| + |BC|$$

FD3D. Continuing from **FD3A**, use **Measure | Calculate** to print

$$\angle A + \angle C \quad \text{and} \quad \angle B + \angle D$$

Confirm that both of these sums equal 180° if and only if the quadrilateral is cyclic.

FD3E. Continuing from **FD3A**, confirm that the incenters of triangles BCD, CDA, DAB, ABC form a rectangle. Sketch the excenters of all four triangles, so that you now have 16 points—do the latest additions determine additional rectangles?

World's Most Famous Theorem

It is often said that the Pythagorean theorem is the world's most famous. Actually, very little is known about Pythagoras himself; and certainly there is no proof that he discovered the theorem. Moreover, it seems likely that Babylonians centuries before Pythagoras were aware of the theorem.

The Pythagorean theorem is the fact that *the square of the longest side-length of a right triangle equals the sum of squares of the other two side-lengths.* The theorem is often abbreviated thus: $c^2 = a^2 + b^2$. Often people get the theorem backwards—what they need for solving certain problems is not the Pythagorean theorem, but rather its converse, namely, that if $c^2 = a^2 + b^2$, then the triangle is a right triangle. For *this* theorem, but not for all theorems, this converse happens to be true, as can be easily verified by the Law of Cosines.

Hundreds of proofs of the Pythagorean theorem are known. We'll discuss three of them, loosely known as the short proof, Euclid's proof, and President Garfield's proof.

The Short Proof

Probably the shortest proof of the Pythagorean theorem is one that uses similar triangles to establish equal ratios. Referring to Figure 7.5, the equal ratios are $b/x = c/b$ and $a/(c-x) = c/a$.

ASSIGNMENT 4.1

FD4A. Emulate Figure 7.5, including labels for segment lengths

$$a \quad b \quad c \quad x \quad c-x$$

Use a two-column format to print a proof, as outlined above, of the Pythagorean theorem. Put numbered statements, such as

$$1. \ \angle DCA = \angle ABC$$

in Column 1 and matching verbal documentation in Column 2. Elsewhere, print measurements of c^2 and $a^2 + b^2$.

FD4B. Continuing, hide all objects except $\triangle ABC$. Then erect squares outwards from the three sides. Measure their areas and confirm that one of them equals the sum of the other two.

FD4C. In **FD4A**, sketch the circles whose diameters are the sides of $\triangle ABC$. Find and confirm a relationship among the areas of these circles.

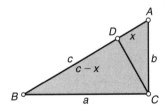

FIGURE 7.5 Short proof of the Pythagorean theorem: Eliminate x from $b/x = c/b$ and $a/(c-x) = c/a$ to get $c^2 = a^2 + b^2$.

Euclid's Proof

Euclid could be described as the head of the mathematics department at the world's first university, at Alexandria, Egypt. Part of the reason for the significance of his principal work, *The Elements,* is its organization of geometry as a purely deductive system, characterized by a short list of basic statements together with a huge body of knowledge deduced from the basic statements. The basic statements are called postulates (or axioms), and what makes them "basic" is that they are *assumed,* rather than proved, because, loosely speaking, no statements that are *more* basic could be found from which they can be deduced.

Euclid's proof of the Pythagorean theorem is a good example of purely deductive method. For present purposes, we won't start all the way back with Euclid's postulates but will instead assume a fair amount of standard background deducible from the postulates.

ASSIGNMENT 4.2

FD4D. Emulate Figure 7.6, including the labels. As indicated, there are three squares, and line *CI* is perpendicular to line *AB*. In a two-column format, as described in **FD4A**, print Euclid's proof, based on the deductions recorded in Figure 7.6.

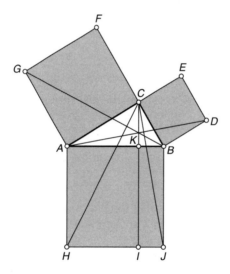

Outline of Euclid's proof of the Pythagorean theorem:

$$\text{Area}(AHJB) = \text{Area}(AHIK) + \text{Area}(KIJB)$$
$$= 2 \cdot \text{Area}(AHC) + 2 \cdot \text{Area}(BCJ)$$
$$= 2 \cdot \text{Area}(ABG) + 2 \cdot \text{Area}(ABD)$$
$$= \text{Area}(CFGA) + \text{Area}(CBDE)$$

FIGURE 7.6 Outline of Euclid's proof

FD4E. Continuing from **FD4D**, add printed numerical values of the sidelengths of $\triangle ABC$ and their squares, a^2, b^2, c^2. Ascertain that c^2 stays equal to $a^2 + b^2$ as you vary the shape of $\triangle ABC$.

President Garfield's Proof

This proof was first published in 1876 in the *New England Journal of Education,* five years before Garfield became the President of the United States. The article opens with these words: "In a personal interview with Gen. James A. Garfield, Member of Congress from Ohio, we were shown the following demonstration of the *pons asinorum,* which he had hit upon in some mathematical amusements and discussions with other M. C.'s". (The Latin *pons asinorum* for *asses' bridge,* dating back to Euclid, means *fools' obstacle*—a bridge certain people will not cross.)

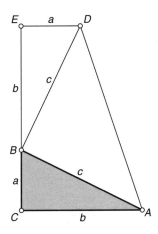

President Garfield's proof:

$$\frac{1}{2}(a + b)(a + b) = \frac{1}{2}ab + \frac{1}{2}ab + \frac{1}{2}c^2$$

$$(a + b)(a + b) = ab + ab + c^2$$
$$(a + b)^2 = 2ab + c^2$$
$$a^2 + b^2 = c^2$$

FIGURE 7.7 President Garfield's proof

In Figure 7.7 trapezoid *ADEC* has base-lengths *a* and *b* and height *a + b.* President Garfield's method depends on expressing the area of the trapezoid in two ways, namely as (height times the average of the base-lengths) and as the sum of areas of three right triangles. Equating these two and simplifying lead to $c^2 = a^2 + b^2$.

ASSIGNMENT 4.3

FD4F. Emulate Figure 7.7. In a two-column format, as described in **FD4A,** present President Garfield's proof. Include measurements of c^2 and $a^2 + b^2$.

FD4G. Starting with $\circ(O, |OZ|)$, sketch a diameter *CA* and let *B* be a movable point on the circle. Extend the right triangle *ABC* as in President Garfield's proof (Figure 7.7). Sketch the loci of points *A, D,* and *E*. Use **Measure | Calculate** to check that $c^2 - a^2 - b^2 = 0$. Animate *B*.

Pascal's Theorem and a Very Useful Tool

While a teenager, Blaise Pascal discovered the following theorem: *if a hexagon is inscribed in a conic section, then the three pairs of the extensions of opposite sides meet on a line.* See Figure 7.8.

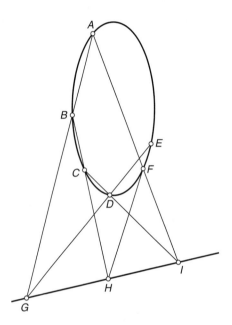

FIGURE 7.8 Pascal's theorem: if A, B, C, D, E, F are on a conic, then G, H, I are collinear

The converse of Pascal's theorem is the fact that if the extensions of opposite sides meet on a line then the hexagon has a circumconic. This converse suggests a very useful tool for sketching conic sections: to construct from 5 given points a 6th point whose locus is the desired conic. Here's the construction, starting with A, B, C, D, E labeled in counterclockwise order—the order can be changed later.

1. Let $G = AB \cap DE$
2. Let E' be a fixed translation of E, specifically a translation that is unlikely to be one of the given points; for example, E' could be 0.56 inches from E in the direction $42°$
3. Let Z be a movable point on the circle centered at E and passing through E'
4. Let $H = BC \cap EZ$
5. Let $I = CD \cap GH$
6. Let $F = EZ \cap AI$
7. Select Z and F, apply **Construct | Locus**, and animate Z

The ellipse in Figure 7.9 was sketched using the above steps.

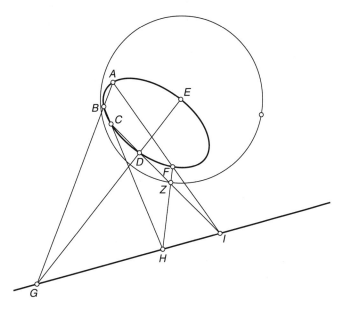

FIGURE 7.9 Converse of Pascal's theorem: if G, H, I are collinear, then A, B, C, D, E, F are on a conic

ASSIGNMENT 5.1

FD5A. Starting with 5 independent points labeled A, B, C, D, E, sketch a point F lying on the conic that passes through the five points. Save as a sketch, including a tool named **conic thr 5 pts**. The only visible result of applying this tool should be the conic.

FD5B. Use **conic thr 5 pts** to sketch the conic that passes through the points $(2, 3)$, $(0, 2)$, $(0, 0)$, $(3, 0)$, $(4, 1)$.

FD5C. Let (x, y) be an independent point. Sketch the conic that passes through the points $(2, 3)$, $(0, 2)$, $(0, 0)$, $(3, 0)$, (x, y). Identify and sketch a region R such that if (x, y) lies in R, then the conic is a hyperbola.

FD5D. Start with 4 independent points labeled A, B, C, D. Let E be a movable point on a circle $\circ(O, |OX|)$. Apply **conic thr 5 pts**, and then animate E—but before doing so, predict the result.

SECTION 6 Brianchon's Theorem

Recall that the dual of a theorem is obtained by interchanging the roles of lines and points. The dual of Pascal's theorem was published by Brianchon in 1810—about 170 years after the discovery of Pascal's theorem.

In order to be very literal about the subject of duality, both Pascal's and Brianchon's theorems are stated here:

PASCAL'S THEOREM. *If a hexagon is inscribed in a conic section, then the three pairs of lines of opposite sides concur in collinear points.*

BRIANCHON'S THEOREM. *If a hexagon is inscribed in a conic, then the lines joining pairs of opposite vertices concur in a point.*

The converse of Brianchon's theorem is also a theorem, and it invites a sketch much in the spirit of **FD5A**. Here, however, instead of sketching the conic passing through 5 given points, we'll construct the conic *tangent* to 5 given *lines*. Call them L_1, L_2, L_3, L_4, L_5. Let

$$A = L_1 \cap L_2 \quad B = L_2 \cap L_3 \quad C = L_3 \cap L_4 \quad D = L_4 \cap L_5$$

Let E be a movable point on L_5, and let $P = AD \cap BE$. By the converse of Brianchon's theorem, if a line through E meets line L_1 in a point F for which the three lines AD, BE, CF concur, then the conic is inscribed in the hexagon. In particular, the line EF is tangent to the conic. The point F, therefore, is constructed as $CP \cap L_1$. By selecting E and line EF, we can apply **Locus** to visualize the conic that is tangent to the given 5 lines. In this interesting use of **Locus**, note that the conic is not plotted directly. Instead, lines are plotted, and the conic appears as their envelope. That is, the conic is formed from the set of lines tangent to it, each of which contributes one point to the conic. Figures 7.10 and 7.11 show an ellipse and a hyperbola obtained in this manner.

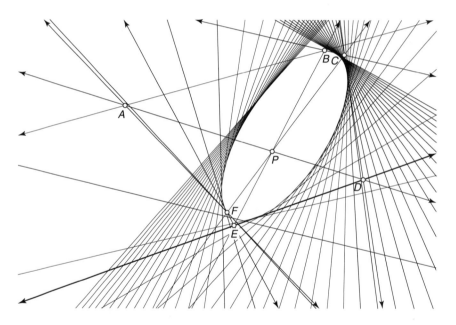

FIGURE 7.10 Tangent lines forming an ellipse. To see that tangent line "go around", drag point E. The tangent lines are produced by selecting point E and line EF and applying **Construct | Locus**.

A Sketchpad realization of this construction will be very sensitive to small changes in the positions of the given five lines.

ASSIGNMENT 6.1

FD6A. Following the discussion just above, emulate Figure 7.10 and save your work as a sketch that includes a tool named **conic from tangents**.

FD6B. Emulate Figure 7.11.

FD6C. Form the pentagon having vertices (2, 3), (0, 2), (0, 0), (3, 0), (4, 1). Use **conic from tangents** to sketch the conic (as an envelope) that is inscribed in the pentagon.

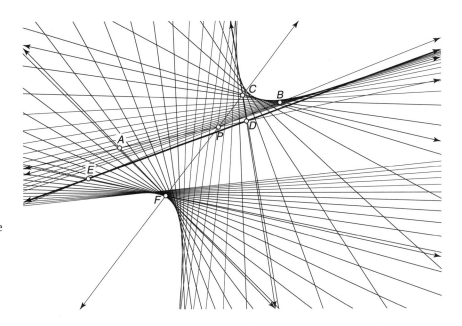

FIGURE 7.11 Tangent lines forming a hyperbola. To see that tangent line "go around", drag point E. The tangent lines are produced by selecting point E and line EF and applying **Construct | Locus**.

SECTION 7 Poncelet Porism and Orbits

Suppose you are given a circle inside a larger circle and wish to construct $\triangle ABC$ for which your two circles are the incircle and the circumcircle—that is, the sides of $\triangle ABC$ are to be tangent to the inner circle, while the vertices lie on the larger circle. Whether such a triangle exists, it turns out, depends on the two circles, but if there is one such triangle, then there are infinitely many. A problem such as this, where there is not just one solution but infinitely many, is called a *porism*. The particular problem posed in this paragraph is possibly the most famous geometric porism. It was discovered by the French geometer, J. V. Poncelet.

Instead of starting with the two circles and constructing $\triangle ABC$, let's start with $\triangle ABC$ and its incircle and circumcircle, as shown in Figure 7.12. We'll construct virtually all the other triangles that share the same incircle, U, and circumcircle, V. Place a movable point t on U. Sketch the tangent line through t and its points of intersection with V, labeled B_t and C_t. Then construct tangent lines to U from B_t and from C_t. These lines meet in a point A_t on V. Triangle $A_t B_t C_t$ has incircle U and circumcircle V. Select point t and circle U and apply **Display | Animate** to see $\triangle A_t B_t C_t$ as a variable triangle that "rotates" through the infinite set of triangles that share the same incircle and circumcircle.

If you now sketch a "point" of $\triangle A_t B_t C_t$, such as the centroid, then when $\triangle A_t B_t C_t$ rotates, the "point" forms a locus, or *orbit*. The word *point* has been put in quotation marks in order to call attention to the fact that the "point" is actually a function, *not* just a point. For example, centroid is a function defined on the domain of all triangles, in the same sense that sine is a function defined on a set of numbers. Calling centroid a point (without the quotation marks) is appropriate when speaking of its location (or "value") in a *particular* triangle, just as $\sin(\pi/3)$ is a value of a function.

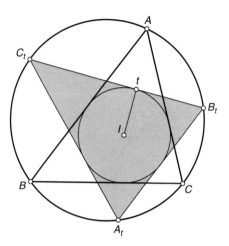

Poncelet porism. As point t "rotates" around the incircle, triangle $A_t B_t C_t$ "rotates" inside the circumcircle.

FIGURE 7.12 Poncelet porism

ASSIGNMENT 7.1

FD7A. Start with $\triangle ABC$, its incircle U and circumcircle V, and sketch the triangle $A_t B_t C_t$ discussed above. Confirm that the area, perimeter, and inradius of $\triangle A_t B_t C_t$ satisfy

$$\text{area/perimeter} = \text{inradius}/2$$

Animate A_t.

FD7B. Continuing from **FD7A**, construct the centroid of $\triangle A_t B_t C_t$ and view its orbit.

FD7C. Continuing from **FD7A**, evaluate each of the following functions at $\triangle A_t B_t C_t$, and view their orbits: orthocenter, midpoint of segment $B_t C_t$, and incenter.

FD7D. If you start with inner and outer circles U and V, then a triangle $\triangle ABC$ having incircle U and circumcircle V exists if and only if $R^2 - d^2 = 2rR$, where r and R are the radii of U and V, respectively, and d is the distance between the centers of U and V. Add to **FD7A** printed calculations of $R^2 - d^2$ and $2Rr$. Do they really stay equal to each other when you vary the shape of $\triangle ABC$?

FD7E. Starting with **FD7A**, add lines tangent to V at B_t and C_t. Let Q be the point of intersection of these tangent lines. Experiment with the locus of Q as a function of t. Does it appear that the locus is always a certain kind of curve encountered elsewhere in this book?

FD7F. Starting with **FD7A**, add the excenters of $\triangle A_t B_t C_t$ and their orbits. Do these orbits have any quickly observable interrelationships?

FD7G. Starting with **FD7A**, sketch the points

$$U = BB_t \cap CC_t \quad V = CC_t \cap AA_t \quad W = AA_t \cap BB_t$$

and sketch their orbits. Do these orbits concur in a point? If so, do you have a tool that sketches this point?

SECTION 8 # Intermediate Triangles

Suppose ABC and $A'B'C'$ are triangles and that segments AA', BB', CC' are shown. Imagine a point a that moves on segment AA' from A to A'. We can use the ratio $|Aa|/|AA'|$ to determine a point b moving on BB' and a point c moving on CC' in such a way that the initial triangle ABC is transformed onto the final triangle $A'B'C'$ by the moving triangle abc. See Figure 7.13. Each such triangle abc can be called an intermediate triangle.

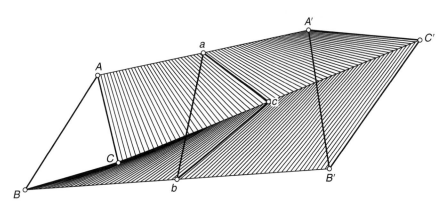

FIGURE 7.13 Intermediate triangles: loci of segments bc and ca, driven by point a

ASSIGNMENT 8.1

FD8A. Emulate Figure 7.13. Include the circumcenter and circumcircle of △abc. Sketch the loci of both of them as functions of a, and animate a.

FD8B. Repeat the construction in **FD8A**, but this time, arrange for △$A'B'C'$ to be similar to △ABC, and confirm using **Measure | Calculate** that the intermediate triangles are similar to △ABC. Someone running your sketch should be able to rotate, translate, and vary the size of △$A'B'C'$ easily, without changing its shape.

FD8C. Starting with **FD8A**, remove the circumcenter and circumcircle, and then move the points A, B, C, A', B', C' so that △$A'B'C'$ is perspective to △ABC with perspector between A and A'. In such a case, must there be a point a for which $a = b = c$?

FD8D. Modify **FD8A** so that △$A'B'C'$ is a single point. Determine experimentally whether the intermediate triangles are similar to △ABC. Figure out the area of the triangle abc in terms of the ratio $|Aa|/|AA'|$, and print measurements that confirm your formula.

Monge's Theorem

Suppose U, V, W are three circles, none touching another. Let X be the point where the external tangents to circles V, W meet. Define Y and Z cyclically. Monge's theorem is, in part, that the points X, Y, Z lie on a line, as depicted in Figure 7.14.

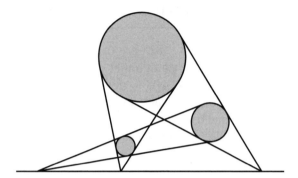

FIGURE 7.14 Monge's Theorem: pairs of external tangents meet in collinear points

If we use internal tangents instead of external, there are three more points, X', Y', Z', and if we combine them with X, Y, Z, there are three additional collinearities. See Figure 7.15.

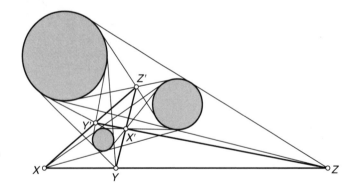

FIGURE 7.15 Extension of Monge's theorem: 3 circles, 12 tangent lines, 4 collinearities

ASSIGNMENT 9.1

FD9A. First, we'll construct external tangents to nonintersecting circles $\circ(A, |AB|)$ and $\circ(C, |CD|)$. Write the radii $|AB|$ and $|CD|$ as R and r, respectively, with $r < R$ and the smaller circle lies outside the larger. Sketch $\circ(A, R - r)$. Let M be the midpoint of segment AC, and sketch $\circ(M, |MA|)$. Let S be a point where $\circ(M, |MA|)$ and $\circ(A, R - r)$ meet. Then line CS is tangent to $\circ(A, R - r)$. Let L be the line through C perpendicular to CS, let T be the point where L meets $\circ(C, r)$ on the same side of line AC as point S. The line through T perpendicular to L is tangent to both given circles. Reflect it about line AC to obtain the other external tangent. Use this sketch to create a tool, **external tangents**. (This tool applies only when the first circle selected is larger than the second. When creating the tool, you may find it necessary to select the points A, B, C, D in order to determine the two circles, instead of selecting the circles themselves.)

FD9B. Here we'll construct an internal tangent. Following the notation of **FD9A**, sketch $\circ(C, R + r)$ and $\circ(M, |MA|)$. Let S be a point where $\circ(M, |MA|)$ and $\circ(C, R + r)$ meet. Then line AS is tangent to $\circ(C, R + r)$. Translate line AS by distance R in the direction of vector SC to obtain an internal tangent line. Reflect it about line AC to obtain the other internal tangent. Use this sketch to create **internal tangents**.

FD9C. Emulate Figure 7.14, using **external tangents**.

FD9D. Add to **FD9C** the points X', Y', Z' where pairs of internal tangents meet. Discover relationships among the points X, Y, Z, X', Y', Z'.

PROJECTS

PROJECT 1: RECIPROCAL PYTHAGOREAN THEOREM

Figure 7.16 shows three right triangles involving sidelengths

$$a = |BC| \quad b = |CA| \quad c = |AB| \quad h = |AD| \quad x = |BD|$$

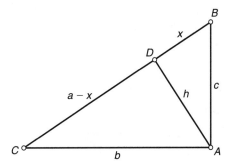

FIGURE 7.16 Right triangles

Part 1. Confirm *and prove* that $h^{-2} = b^{-2} + c^{-2}$. This is sometimes called the *reciprocal Pythagorean theorem*.

Part 2. Create a sketch with printed measurements that confirm that $h^{-2} = b^{-2} + c^{-2}$ for thousands of right triangles that share a common circumcircle.

PROJECT 2: QUADRILATERALS AND CIRCLES

Part 1. Confirm and prove, as an easy consequence of Ptolemy's theorem (Figure 7.3), that if ABC is an equilateral triangle and P lies on the arc BC of the circumcircle of ABC, then $|PA| = |PB| + |PC|$.

Part 2. Seek, confirm, and prove a result that relates to Figure 7.4, in the case of an equilateral triangle ABC, in the manner that **Part 1** relates to Figure 7.3.

Part 3. In the configuration of Figure 7.3, sketch lines tangent to the circle at points A, B, C, D. Label the vertices of the resulting quadrilateral counterclockwise as A', B', C', D', with A' being the first of these after A, counterclockwise. Determine, with a sketch, which of the quadrilaterals

$$AA'BO \qquad BB'CO \qquad CC'DO \qquad DD'AO$$

can be inscribed in a circle.

Part 4. In the configuration of **Part 3**, determine which of the quadrilaterals

$$AA'BO \qquad BB'CO \qquad CC'DO \qquad DD'AO$$

can be circumscribed about a circle.

PROJECT 3: A JAPANESE TEMPLE THEOREM

In *Modern Geometry,* Roger A. Johnson writes that "it was the ancient custom of Japanese mathematicians to inscribe their discoveries on tablets which were hung in temples, to the glory of the gods and the honor of the authors". One such theorem hanging in 1800 is stated here:

Suppose points P_1, P_2, ..., P_n are arranged in order on a circle. Draw the segments

$$P_1 P_2 \qquad P_2 P_3 \qquad \ldots \qquad P_{n-1} P_n \qquad P_n P_1$$

Pick any one of the P_i, and draw segment $P_i P_j$ for each $j \neq i$. (See Figure 7.17.) Inscribe a circle in each of the resulting triangles (as in Figure 7.18). Then the sum of radii of the circles is independent of i.

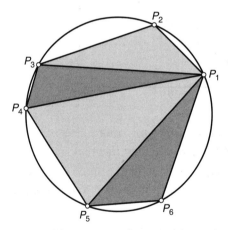

FIGURE 7.17 Points on a circle with triangulations from point P_1

That is, you get the same sum, no matter which of the points you choose for a common vertex of the triangles.

Illustrate the Japanese temple theorem using $n = 6$ in the following way. Sketch a circle, place movable points P_1, P_2, ..., P_6 on it, and translate, so that you have two identical and well separated configurations. On the first circle, triangulate from P_1, and on the second, from P_2. Print the sums of radii for both configurations.

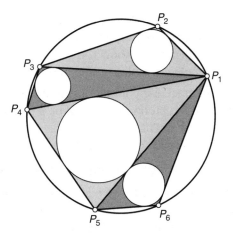

FIGURE 7.18 Inscribed circles

PROJECT 4: PAPPUS'S THEOREM

A statement of this theorem, attributed to an ancient Greek geometer, follows. Suppose M and N are lines, points A, B, C lie on M, and points D, E, F lie on N. Let

$$G = BF \cap CE \quad H = AF \cap CD \quad I = AE \cap BD$$

Then the points G, H, I are collinear.

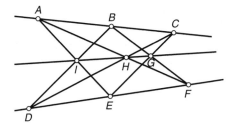

FIGURE 7.19 Pappus's theorem: collinear points of intersection

Part 1. Confirm Pappus's theorem using Sketchpad. If point B is moved to the other side of point C, do the points G, H, I remain collinear?

Part 2. Discover conditions under which lines AB, DE, GH concur in a point.

PROJECT 5: INTERMEDIATE OBJECTS

Part 1. Continuing **FD8B**, in which the triangles ABC and $A'B'C'$ are similar, figure out and confirm a formula for the perimeter of $\triangle abc$ in terms of the ratio $r = |Aa|/|AA'|$ in the case that the lines AA', BB', CC' concur in a point.

Part 2. Looking at all the lines and proportional distances in Figure 7.13, someone might expect that the sidelength $|bc|$ varies linearly from $|BC|$ to $|B'C'|$ as a varies from A to A'. Verify with an appropriate *curved* locus that this function is not linear.

Part 3. Sketches **FD8A–FD8D** suggest an intriguing possibility: to sketch an object $\mathcal{H}_L$ on the left and another, $\mathcal{H}_R$ on the right, and then to sketch intermediate objects that show the continuous transformation of $\mathcal{H}_L$ into $\mathcal{H}_R$. Apply this idea to a pair of humanoids.

Part 4. The intermediate humanoids of **Part 3** depend on **Transform | Dilate**. Other intermediate humanoids could be generated using **Transform | Rotate**. Fulfill this prospect. Further actions may be investigated by applying **Edit | Merge Point to Circle** to knees and elbows.

PROJECT 6: SLOW-MOTION TRANSFORMATIONS

In Project 5, line segments are continuously dilated from an initial position to a terminal position. This technique can be applied to show continuous (or "slow-motion" transformation) from one curve to another. In this project, such transformations include inversion in a circle and isogonal conjugation.

Part 1. Start with a circle $\circ O$ and a line L that does not touch $\circ O$. Apply the tool **inverse in line** to sketch the inverse of L in $\circ O$. Place several points on L and invert each one. Sketch the segment from each point to its inverse. Place a movable point X on one of these segments, with endpoints labeled P and Q. Use the ratio $r = |QX|/|QP|$ to obtain intermediate points on the other segments. Join successive pairs of the intermediate points to form a polygonal curve. This segment approximates a curve that could be described as "r of the way" between the line and its inverse (which is a circle). When you animate X, this moving curve suggests a slow-motion transformation L into its inverse and, in reverse, from the inverse back to L. Be sure to vary the positioning of L and the center and radius of $\circ O$ during animation. See Figure 7.20.

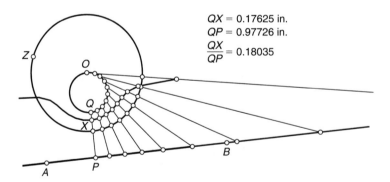

$QX = 0.17625$ in.
$QP = 0.97726$ in.
$\dfrac{QX}{QP} = 0.18035$

FIGURE 7.20 Slow-motion inversion of a line in a circle

Vary the line AB by dragging A and B.
Vary circle (O, Z) by dragging O and Z.

Part 2. Create a tool to be named **inverse of circle** that inverts a circle in a circle in the manner in which **inverse in line** inverts a line in a circle.

Part 3. Start with a circle $\circ O$ and a circle U that does not touch $\circ O$. Apply the tool **inverse in circle** to sketch the inverse of U in $\circ O$. Place several points on U and invert each one. Continue, using the method in **Part 1**, to create a sketch of "slow-motion" inversion between U and its inverse in $\circ O$.

Part 4. Apply the same method to show "slow-motion" isogonal conjugation of a line L with respect to a triangle ABC. (Recall that the image of L is an ellipse, parabola, or hyperbola accordingly as the number of points in which L meets the circumcircle of $\triangle ABC$ is 0, 1, or 2.)

C H A P T E R **E I G H T**

Selected Topics

THIS CHAPTER BEGINS with simultaneous rotation and translation—two actions well performed by Sketchpad—as the seed for the topic of polar coordinates. Simple combinations of rotation and translation produce spirals. More liberal combinations produce roses and cardioids. These and other curves are important enough that Sketchpad offers built-in polar graphing, so that we needn't literally put such graphs together as simultaneous rotation and translation. It is, however, sometimes very helpful to realize that polar graphing is basically such a simple combination.

Likewise, graphing in the xy-plane is a combination of two transformations, both of them being translations. Thus, the two kinds of graphing, polar and cartesian, have this much in common: they consist of transformations of a single point, the origin.

There are certain curves that are much easier to represent and graph in polar coordinates than in cartesian. One of our objectives is to gain an understanding of these curves.

After sampling Sketchpad's polar graphing capabilities, we'll turn to a problem that starts with any five points, no three of which are collinear. We know from Section 5 of Chapter 7 that the five points determine a conic that passes through them. Here in Chapter 8 we'll construct the center of that conic.

Certain numbers are not <u>ratios</u> of integers and so deserve the name *ir*<u>*ratio*</u>*nal*. These include

$$\sqrt{2} \qquad \sqrt{3} \qquad 2^{1/3} \qquad \pi \qquad e$$

Another frontrunner among irrational numbers is $(1 + \sqrt{5})/2$, often called the *golden mean*. We'll take a look at geometric constructions involving this special number.

Next, we'll investigate the possibility of inscribing a similar copy of a given triangle in another given triangle. This, it turns out, can be done in infinitely many ways.

Sketchpad enables you to tessellate a region by repeated applications of **Transform | Translate**. In Section 7, we'll see that many shapes can be turned into "wallpaper" in this way. A fundamental geometric question underlying this subject is this: what shapes tessellate? (Some do and some don't.)

SECTION 1   Spirals

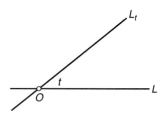

FIGURE 8.1 Line L_t obtained by rotating line L

Here is a thought experiment. Imagine a point O that will stay fixed and another point P that will start at O and move away according to a certain rule. To describe the rule, let L be a line through O. If you wish, think of L as the x-axis and O as the origin. Now suppose we create a new line L_t by rotating L about O; specifically, L_t is the rotation of L exactly t radians counterclockwise. For example, if L is the x-axis, then $L_{\pi/2}$ is the y-axis. In Figure 8.1, the angle t is between $\pi/6$ and $\pi/4$.

Now ask: *where is the point P on L_t whose distance from O is t?* When $t = 0$, the easy answer is that $P = O$. As angle t increases, P moves away from O: the bigger the angle, the bigger the distance. In your mind, you should see a spiral. Let's use Sketchpad to confirm this expectation.

First, since we'll be using the angle as a distance, we must measure it in radians, not degrees. Sketch an xy-coordinate system with origin labeled O and the unit point on the x-axis labeled U. Sketch the circle having center O and passing through U. Put a movable point s_t on the circle, and sketch ray Os_t. Measure the arclength from U counterclockwise to s_t; call it r, and then sketch the point P on ray Os_t at distance r from O. Finally, sketch the locus of P as driven by s_t, as shown in Figure 8.2.

The spiral in Figure 8.2 is somewhat limited because Sketchpad's domain for arclength is restricted. Specifically, the arclength t extends from $-\pi$ to π. To see more of spirals than one wrap-around, we shall place the underlying variable t on the x-axis, instead of a circle, and apply domain-magnification, as described on page 93. The main idea is to use the point $(\cos t, \sin t)$, since this point provides a central angle of t radians. Details are given in the caption of Figure 8.3.

ASSIGNMENT 1.1

ST1A. In an xy-coordinate system, sketch a spiral as the locus of a point P' that is a function of a movable point P on a circle centered at the origin. As always, print your observations.

ST1B. Sketch a spiral as the locus of a point P' that is a function of a movable point t on the x-axis, using sliders as indicated in Figure 8.3.

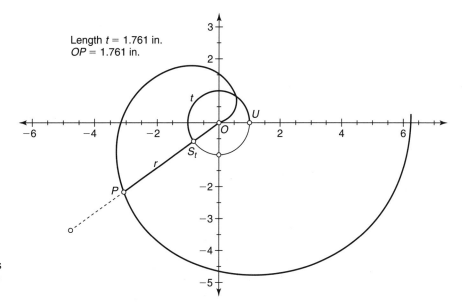

Length $t = 1.761$ in.
$OP = 1.761$ in.

FIGURE 8.2 Portion of the spiral $r = t$, where t is the length of segment OP and is also an arclength. Drag s_t to vary t and r.

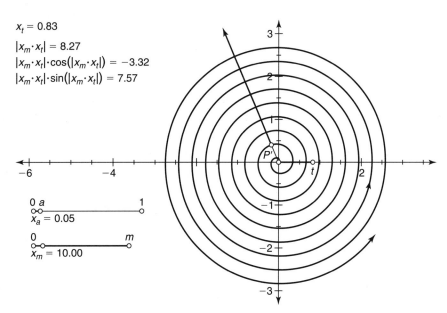

$x_t = 0.83$

$|x_m \cdot x_t| = 8.27$
$|x_m \cdot x_t| \cdot \cos(|x_m \cdot x_t|) = -3.32$
$|x_m \cdot x_t| \cdot \sin(|x_m \cdot x_t|) = 7.57$

$0\ a \qquad\qquad 1$
$x_a = 0.05$

$0 \qquad\qquad m$
$x_m = 10.00$

FIGURE 8.3 The spiral $r = a \cdot t$ for $t > 0$. Drag a to vary the density of the spiral. Drag m to vary the domain of $r = r(t)$. Drag t to vary P'.

SECTION 2 Polar Coordinates I

One of the two polar coordinates, namely the angle, is traditionally denoted by the Greek letter θ. Until now, we've been using the symbol t for this angle, this being a standard symbol in the subject of parametric equations, especially when the parameter is *time,* as in the study of motion. Now we'll switch to θ, as this is the symbol used in Sketchpad's polar coordinate grapher.

Recall from Section 1 that for each angle θ of rotation about a center, O, a point P is separated from O by a distance that depends on θ. This distance, $|OP|$, is therefore a function of θ. We shall abbreviate $|OP|$ as r and write $r = r(\theta)$ to indicate that r really is a function of θ. This functional dependence is the key idea for polar coordinates: every point P in the plane has a name (r, θ) given by

$$r = \text{distance from } O \text{ to } P \text{ in the direction } \theta$$

Here are some details and suggestions:

1. The fixed point O is called the *pole* (or origin). If P is any point except the pole, then an angle θ for P is the angle made by swinging the positive x-ray counterclockwise until it passes through P. There are infinitely many other angles for P, too, such as $\theta + 2\pi$, $\theta + 4\pi$, and $\theta - 24\pi$, so that, unlike xy-coordinates, in polar coordinates, every point has more than one representation.

2. Strictly speaking, when we write $r = r(\theta)$, we're referring to a function named r, defined as the set

$$\{(r, \theta) : r = r(\theta) \text{ and } \theta \in D\}$$

where D is a prescribed (or tacitly understood) set called the domain of r. The "graph of r", or more common but less accurate, the "graph of $r(\theta)$", means a picture representing the set of ordered pairs. It is far from possible to show *all* the ordered pairs. Think of it this way: there are infinitely many points (r, θ) as θ varies from 0 to 2π, but there are only finitely many atoms on Earth, so there's not enough matter available. Indeed, not only are atoms much, much too big to be geometric points, but also, atoms are much, much too shaky. Accordingly, a graph, as a representation of a function, isn't "the real thing", but rather a model to assist our understanding of "the real thing".

3. The angle θ may be measured in degrees or radians. However, if in a problem θ plays a role as both an angle and a distance, then θ must be measured in radians. Recall that 1 radian measures the angle subtended by an arc of length 1 on the circle $x^2 + y^2 = 1$. Conversions are therefore given by

$$1 \text{ radian} = 180/\pi \text{ degrees, so that}$$

$$\theta \text{ radians} = 180\theta/\pi \text{ degrees}$$

consequently,

$$1 \text{ degree} = \pi/180 \text{ radians, so that}$$

$$\theta \text{ degrees} = \pi\theta/180 \text{ radians}$$

Sometimes is it helpful to convert from (r, θ) coordinates to xy-coordinates, or the reverse. Conversion is carried out by substitutions from the basic identities:

$$x = r\cos\theta \qquad y = r\sin\theta$$

From these two, we have $r^2 = x^2 + y^2$ and $\tan\theta = y/x$.

For example, the line $y = ax$ yields $y/x = a = \tan\theta$, so that $\theta = \arctan a$. Since a, the slope of the line, is constant, we conclude that any nonvertical line through O has a very simple polar equation, namely, $\theta = $ a constant. The vertical line through O fits the scheme well, too: $\theta = \pi/2$.

Here's another example, this time starting with $r = 4\cos\theta$. Multiply both sides by r and then substitute $x^2 + y^2$ for r^2 and x for $r\cos\theta$. Cancel r and have $x^2 + y^2 = 4x$. If at first you don't recognize this as a familiar curve, complete a square to put the equation into the standard form

$$(x - h)^2 + (y - k)^2 = \rho^2$$

Among curves which have polar representations much nicer than their xy representations are spirals, with equations as simple as $r = \theta$. (Try converting *that* to xy-coordinates and solving for y in terms of x!)

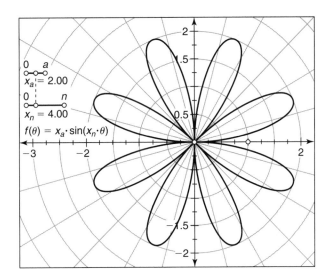

FIGURE 8.4 The rose $r = 2\sin 4\theta$. Drag a to vary the petal size. Drag n to vary the number of petals.

ASSIGNMENT 2.1

ST2A. Graph together the following curves given by equations in polar coordinates, after applying **Edit | Preferences | Units | Angle | radians**:

$$r = 1 \quad r = 2 \quad r = 3 \quad r = \theta$$

Why does the graph of $r = \theta$ "stop", and what could you do to "keep it going"? As always, print your answers in captions.

ST2B. Graph together the following curves:

$$r = \theta \quad r = 2\theta \quad r = \sqrt{\theta}$$

ST2C. Graph together the following curves:

$$r = \cos\theta \quad r = 2\cos\theta \quad r = 3\cos\theta$$

Then print equations for these curves using xy-coordinates.

ST2D. Graph together the following curves:

$$r = \sin\theta \quad r = 2\sin\theta \quad r = 3\cos\theta$$

Then print equations for these curves using xy-coordinates.

ST2E. Graph together the six graphs given by $r = a\cos b\theta$, for

$$a = 1 \quad a = 2 \quad b = 1 \quad b = 3 \quad b = 5$$

Predict features of the graph of $r = 3\cos 7\theta$. Use Sketchpad to confirm or refute your predictions.

ST2F. Curves of the form $r = a\cos b\theta$ and $r = a\sin b\theta$ comprise a family known as roses. Graph together the four roses given by $r = a\sin b\theta$, for

$$a = -3 \quad a = 2 \quad b = 2 \quad b = 4$$

Predict features of the graph of $r = -4\sin 6\theta$. Use Sketchpad to confirm or refute your predictions.

ST2G. Generalize the results of **ST2E** by starting with sliders for a and b, so that you can watch how the graph depends continuously on changes in a and b. (Sliders are introduced in connection with Figure 3.13.)

SECTION 3 # Polar Coordinates II

Aside from spirals, other simple curves whose polar representations are more amenable than rectangular representations are limaçons, as in Figure 8.5.

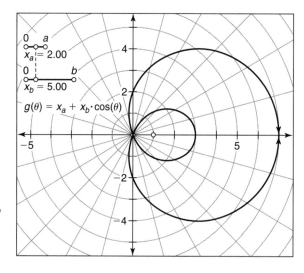

FIGURE 8.5 The limaçon $r = a + b\cos\theta$. Drag a and b to vary the size and shape. The shape is determined by the ratio a/b.

ST3A. Graph together the following curves:

$$r = 1 + \cos\theta \quad r = 2 + \cos\theta \quad r = 2 + 3\cos\theta$$

These curves exemplify a *cardioid*, a *dimpled limaçon*, and a *limaçon with inner loop*, respectively.

ST3B. Graph together the following curves:

$$r = 1 + \sin\theta \quad r = 2 + \sin\theta \quad r = 2 + 3\cos\theta$$

These curves exemplify a *cardioid*, a *dimpled limaçon*, and a *limaçon with inner loop*, respectively.

ST3C. Generalize the results of **ST3A** by starting with sliders for a and b, so that you can watch how the graph of

$$r = a + b\cos t$$

depends continuously on changes in a and b.

ST3D. Add to **ST3C** graphs of the circles

$$r = a \quad \text{and} \quad r = b\cos t$$

Then explain in a caption how the graph of $r = a + b\cos t$ can be regarded as a "geometric sum" of the two circles.

ST3E. Use **FD7B** (Figure 7.12) to sketch the orbits of the midpoints of the sides of $\triangle A_t B_t C_t$. What connection does that orbit have with the present topic of polar coordinates?

SECTION 4   Conic Center of Five Points

Recall that 2 points determine a line, 3 a circle, and 5 a conic. Earlier the conic through 5 points was sketched. Now we ask how to construct the center of that conic. Of course, if any three of the points lie on a line, then the conic is simply that line; in this case, the conic is called *degenerate* and has no center. (Well, that's not quite right: actually, its center is the point where all perpendiculars to the conic-line meet the line at infinity). We shall assume that no 3 of 5 given points A, B, C, D, E are collinear.

Figure 7.8 shows a choice of A, B, C, D, E and Section 5 of Chapter 7 explains how their conic can be constructed as the locus of a sixth point, shown as point F in Figure 7.8. Loosely speaking, the point F is constructed so that the points G, H, I are collinear, and this implies, by the converse of Pascal's theorem, that F lies on the conic determined by A, B, C, D, E.

The key to locating the center of the conic is the notion of "parallel chords". A chord of a conic is like that of a circle: a segment whose endpoints lie on the conic. It is well known that the line joining the midpoints of parallel chords passes through the center of the conic—this is no surprise—just pencil a quick sketch. It follows that the point of intersection of two such lines is the center of the conic.

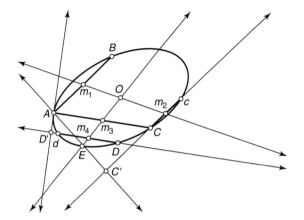

FIGURE 8.6 Center, O, of the conic of five points A, B, C, D, E, constructed as the point of intersection of lines joining midpoints of chords

In order to carry out the sketch, we'll need a tool that deserves to be named **point on conic**. Here's what it must do: given 5 points T, U, V, W, X that determine a conic Ψ, and given a point P not on Ψ, sketch the point t, other than T, where line PT meets Ψ. To construct t, let

$$L = UV \cap WT \qquad M = VX \cap TP \qquad N = XW \cap LM$$

Then, as indicated by Figure 8.7,

$$t = TP \cap UN$$

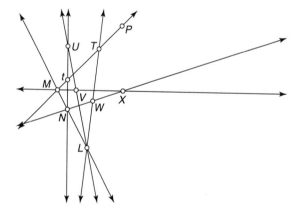

FIGURE 8.7 Construction of point t, where line PT meets the conic of points T, U, V, W, X

Now, given 5 points A, B, C, D, E, we are ready to construct the center of the conic which passes through the 5 points. You can sketch the conic using **conic thr 5 pts** first, if you wish, but, of course, Sketchpad cannot use the conic itself in a construction. It will, nevertheless, be helpful to have a symbol for the conic, and we'll continue to use the symbol Ψ.

1. Sketch m_1 = midpoint of segment AB

2. Sketch line ℓ_1 through C parallel to line AB

3. Sketch C' = the foot of the perpendicular from A onto line ℓ_1
4. Sketch $c = \ell_1 \cap \Psi$ by applying **point on conic** to C, A, B, D, E, C' in that order
5. Sketch m_2 = midpoint of segment Cc
6. Sketch m_3 = midpoint of segment AC
7. Sketch line ℓ_2 through D parallel to line AC
8. Sketch D' = the foot of the perpendicular from A onto line ℓ_2
9. Sketch $d = \ell_2 \cap \Psi$ by applying **point on conic** to D, A, B, C, E, D' in that order
10. Sketch m_4 = midpoint of segment Dd
11. Sketch $O = m_1 m_3 \cap m_3 m_4$

Point O is the required center of Ψ.

ASSIGNMENT 4.1

ST4A. Construct the center O described in this section. Save the result as a sketch, including a tool **conic center**, for which the givens are A, B, C, D, E and the result is O.

ST4B. Apply **conic center** to the five points A_t, B_t, C_t, G_t, H_t, where $\triangle A_t B_t C_t$ is the rotating triangle in the Poncelet porism, and

$$G_t = \text{centroid of } \triangle A_t B_t C_t \quad H_t = \text{orthocenter of } \triangle A_t B_t C_t$$

Does the resulting center O_t appear to lie on any particular object? (The Poncelet porism is illustrated by Figure 7.12.)

ST4C. On a circle with center O, sketch movable points A and B, and sketch the diameters AA' and BB'. Let C be an independent point outside the circle. Apply **conic thr 5 pts** and **conic center** to the points A, B, A', B', C. Print what you observe about the center of the conic when the points A and B are dragged.

ST4D. Figure out how to construct the line tangent to conic $ABCDE$ at a point T on the conic. Then develop a sketch that fulfills your figuring. Include the creation of a tool named **conic tangent**.

SECTION 5   # The Golden Mean

Consider a rectangle shaped so that if you remove the biggest possible square from one end, the remaining rectangle will have the same shape as the original. Figure 8.8 shows a rectangle $ABCD$ that has the right shape.

Any two rectangles having this special shape are similar to each other, so all such rectangles, except for size, are the "same". (In case you've studied equivalence classes, all these rectangles comprise one class.) So, in a sense, we're talking about a single rectangle, which for centuries has

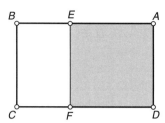

FIGURE 8.8 Square *AEFD* is removed from rectangle *ABCD*, leaving rectangle *EBCF* having the same shape as rectangle *ABCD*

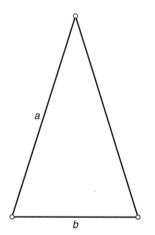

FIGURE 8.9 5-pointed star

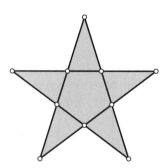

FIGURE 8.10 Triangle cut from the 5-pointed star

been called "the golden rectangle". According to folklore, this rectangle is important in advertising, architecture, and nature, and there is a basis for some of this lore—for more, see George Markowsky, "Misconceptions about the Golden Ratio", *College Mathematics Journal* 23 (1992) 2-19.

Let's apply a bit of algebra to the golden rectangle: its shape is given by the ratio L/W, where L = length and W = width. With reference to Figure 8.8, the requirement that rectangle *EBCF* have the shape L/W takes the form of an equation:

$$\frac{W}{L - W} = \frac{L}{W}$$

which readily simplifies to $L^2 - LW - W^2 = 0$, hence to $x^2 - x - 1 = 0$, where $x = L/W$. By the quadratic formula, $x = (1 + \sqrt{5})/2$. That is, the shape of the golden rectangle is given by "length equals (width times the golden mean)". The number $(1 + \sqrt{5})/2$ is called not only the *golden mean,* but also the *golden ratio* and *golden section*. Unfortunately, these names are also used for the ratio W/L, which equals $L/W - 1$. In this book, "golden mean" means the larger of the two, approximately 1.6180339887498948482.

A second geometric introduction to the golden mean asks for "the point that divides a line segment internally and externally in the same ratio". What this means for segment *CD* in Figure 8.8 is the point F, which divides *CD* so that the external ratio $|CD|/|FD|$ equals the internal ratio $|FD|/|FC|$. These two ratios are simply L/W and $W/(L - W)$, which are equal when L/W is the golden mean, as proved above.

We turn now to a third construction involving the golden mean, beginning with a 5-pointed star, as in Figure 8.9.

This star includes five congruent isosceles triangles, each of which has the shape indicated in Figure 8.10. You should verify that the angles of this triangle measure 36°, 36°, and 72°.

Using the fact that $\sin 18° = (\sqrt{5} - 1)/4$, we have

$$\frac{b/2}{a} = \frac{\sqrt{5} - 1}{4}$$

which implies that $a/b = (1 + \sqrt{5})/2$. So, if you can construct a regular pentagon and a segment of length 1, then you can construct a segment whose length is the golden mean. Figure 8.11 offers yet another construction for the golden mean.

ASSIGNMENT 5.1

ST5A. Construct a 5-pointed star as in Figure 8.9.

ST5B. Start with a segment about one inch long with endpoints labeled D and F. Then construct a point C on line *DF* satisfying

$$|DC|/|DF| = \text{golden mean}$$

ST5C. Let D and E be the midpoints of the sides *AB* and *AC* of an equilateral triangle. Extend *DE* to meet the circumcircle of $\triangle ABC$ and F.

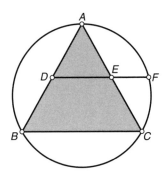

FIGURE 8.11 *ABC* is an equilateral triangle, *D* and *E* are midpoints, and $|DF|/|DE|$ = golden mean

Confirm using Sketchpad that

$$\frac{|DF|}{|DE|} = \frac{|DE|}{|EF|}$$

Then write out an algebraic proof that $|DF|/|DE|$ = golden mean.*

ST5D. Sketch a regular pentagon *ABCDE* and diagonals *AD* and *EC*. Let $Q = AD \cap EC$. Confirm that

$$\frac{|AD|}{|AQ|} = \frac{|AQ|}{|QD|}$$

Then *prove* that $|AD|/|AQ|$ = golden mean.

SECTION 6 # Inscribed Similar Triangles

Imagine two arbitrary triangles: can a copy of one be inscribed in the other? More precisely, if the triangles are labeled as *ABC* and *DEF*, is there a triangle *D'E'F'* similar to *DEF* such that *D'*, *E'*, *F'* lie on sidelines *BC*, *CA*, *AB*, respectively? The affirmative answer to this question sparks many bright fires; in fact, for every possible shape of triangle, there are *infinitely* many triangles of that same shape that are inscribed in any given triangle. Figure 8.12 shows a simple construction of *D'E'F'* similar to *DEF* and inscribed in *ABC*.

FIGURE 8.12 *ABC* and *DEF* are given triangles. To inscribe a triangle similar to *DEF* in *ABC*, place a point *d'* on line *BC* and translate *DEF* by vector *Dd'*. Let $f' = fd' \cap AB$. The line through *f'* parallel to line *fe* meets line *ed'* in a point *e'*. Let $E' = Be' \cap CA$. Let *D'* be the point where the line through *E'* parallel to *d'e'* meets line *BC*, and let *F'* be the point where the line through *E'* parallel to *e'f'* meets line *AB*. Triangle *D'E'F'* is then clearly similar to *DEF* and inscribed in *ABC*.

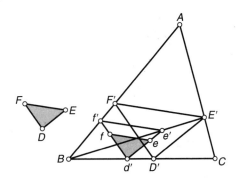

Now that we know that a triangle of any shape can be inscribed in any given triangle, we are ready to account for *all* the triangles of that shape that can be inscribed in a triangle. The key is that we can start with any one of them and "twirl" it through all the others. This remarkable action depends on a particular point *P* that serves as a center about which the family of similar triangles twirls. (By "twirl" we mean "rotate and change size but not shape".) To construct *P* we use a theorem which Roger Johnson in *Modern Geometry* attributes to an 1838 publication by Auguste Miquel.

---

*George Odom, Problem E3007, *American Mathematical Monthly*.

Miquel's theorem states that *if DEF is a triangle inscribed in ABC (as in Figure 8.13), then the circles through A, F, E, through B, D, F, and through C, E, D concur in a point.*

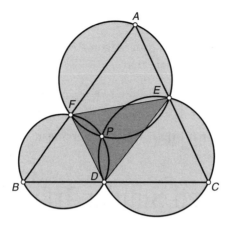

FIGURE 8.13 Miquel's theorem: the point *P* common to three circles

A formula for the "Miquel point" *P* is given in terms of angles by

$$\angle BPC = \angle BAC + \angle EDF$$

and a consequence of this formula is that *P* determines three equal angles as shown in Figure 8.14.

$m\angle PEA = 94.4°$
$m\angle PFB = 94.4°$
$m\angle PDC = 94.4°$

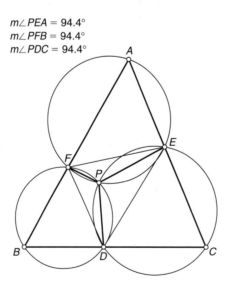

FIGURE 8.14 Point *P* makes equal angles *PEA*, *PFB*, *PDC*

Let's call the equal angles the *inclinations* determined by the inscribed triangle *DEF*. You can imagine moving *D* a bit, and then adjusting *E* and *F* in such a way that the inclinations stay equal, but not the same as for *DEF*. Thus, we may think of an inscribed triangle *D'E'F'* which varies in such a way that the inclinations stay equal to each other as *D'* traverses line *BC*. This variable triangle ranges through all the inscribed triangles similar to *DEF*.

(It is understood that the sides of *ABC* are extended, and that "inscribed" means that *D*, *E*, *F* are on lines *BC*, *CA*, *AB*, respectively.)

Proofs of Miquel's theorem and related theorems are given in Johnson's *Modern Geometry*. We forego proofs here, as our main objective is a construction using the following steps:

1. Start with an arbitrary △*ABC*. Be sure to work with *lines BC, CA, AB*, not segments. Place a movable point *D* on sideline *BC*. Place a movable point *P* inside △*ABC*. (Later, *P* can be dragged outside △*ABC*)

2. Sketch *B′* = reflection of *B* in *C*

3. Sketch *X* = rotation of *D* about *P* through ∠*B′CA*

4. Sketch *E* = *PX* ∩ *CA*

5. Sketch *C′* = reflection of *C* about *A*

6. Sketch *Y* = rotation of *E* about *P* through ∠*C′AB*

7. Sketch *F* = *PY* ∩ *AB*

8. Triangle *DEF* is the desired triangle. By dragging *D*, you can see △*DEF* twirl through virtually all inscribed triangles similar to the original △*DEF*. Figure 8.15 shows five positions of the variable triangle *DEF*.

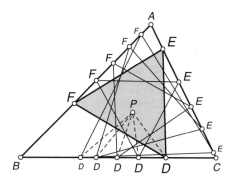

FIGURE 8.15 The variable triangle *DEF*

ASSIGNMENT 6.1

ST6A. Emulate Figure 8.15, showing only one triangle *DEF*. Print measurements of the three equal angles, as well as these: ∠*FDE*, ∠*DEF*, ∠*EFD*, and ascertain that these stay unchanged when you drag point *D*.

ST6B. Using **ST6A**, determine what value of the equal angles corresponds to the triangle *DEF* that has minimal area. Peruse Chaper 2, if necessary, to recognize this triangle as one of a certain well-known family of triangles.

ST6C. The three circles in Figure 8.14 are known as the Miquel circles of *P*. Confirm that their centers are vertices of a triangle similar to △*ABC*.

ST6D. Confirm the following assertion using Sketchpad. Suppose three circles concur in a point, and *A* is an arbitrary point on one of the circles. Then there exists a triangle with vertices on the circles, whose sides pass through the pairwise intersections of the circles, and *A* is one of the vertices. (Technological limitations may make it necessary to restrict *A* to a portion of its circle.)

ST6E. Continuing from **ST6D**, confirm that as the point *A* is dragged, the resulting triangles stay mutually similar.

ST6F. Confirm the following theorem, which is equivalent to Miquel's theorem: if three circles *ABF*, *BCD*, *CAE* concur in a point, then the circles *AEF*, *BFD*, *CDE* concur in a point. (The designation "circle *XYZ*" means the circle that passes through the points *X*, *Y*, and *Z*.)

ST6G. Confirm with Sketchpad the following proposition, known as Johnson's theorem: *if three congruent circles concur in a point, then their other three intersections lie on a fourth congruent circle.*

SECTION 7 Tessellations

Let's start with the chessboard of Figure 1.14. Using translations, you could extend the pattern of squares to cover the screen, or conceptually, to cover the whole plane. However, you wouldn't be able to do that with a regular pentagon as the fundamental shape. That is, squares *tessellate,* and regular pentagons don't. The observation prompts a question: *for what numbers n does the regular n-gon tessellate?* The answer is easy: $n = 3, 4, 6$. That these work and no others do is surprisingly easy to prove. Indeed, at each vertex in such a tessellation, suppose we want to have *m* polygons. At each vertex, every angle has measure $(n - 2)\pi/n$. Therefore, going all the way around such a vertex gives

$$m(n - 2)\pi/n = 2\pi$$

which is equivalent to

$$(m - 2)(n - 2) = 4$$

This equation has only three solutions using integers greater than 2. These solutions (m, n) are $(3, 6)$, $(4, 4)$, and $(6, 3)$, corresponding to the following configurations at each vertex:

- 3 regular hexagons
- 4 squares
- 6 equilateral triangles

You've seen these in various kinds of tiling on floors and walls, and you could easily create similar Sketchpad tessellations. But you've also seen more elaborate tessellations, arising from fundamental shapes other than regular

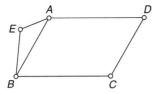

FIGURE 8.16 Step 2

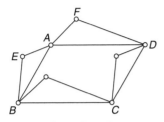

FIGURE 8.17 Step 4

polygons. Among the most popular are those that start with a parallelogram. A step-by-step development for creating such a tessellation follows:

1. Let *ABCD* be the vertices, in counterclockwise order, of a parallelogram.
2. Let *E* be an independent point as indicated in Figure 8.16.
3. Translate △*ABE* by vector *AD*.
4. Let *F* be an independent point as indicated in Figure 8.17.
5. Translate △*ADF* by vector *AB*.
6. Your vertices now determine a polygon that tessellates; color it.
7. Translate your polygon (vertices, edges, and interior) by vector *AD*.
8. Translate all objects except segments *AB*, *BC*, *CD*, *DA* by vector *AB*.
9. You now have four images of the original polygon; use a second color for polygons having a common edge, as indicated in Figure 8.18. Use **Translate** to continue tessellating. (The cover design of this book was generated in this way.)

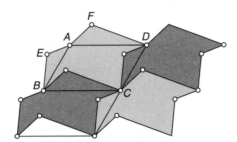

FIGURE 8.18 Step 9

Some very attractive tessellations can be produced by using several points in place of the single point *E* and several points in place of the single point *F*. You can find websites that show such tessellations, in which the fundamental shapes (based on a hidden parallelogram) resemble birds, lizards, and insects. You may have seen related designs by the master tessellator, M. C. Escher.

ASSIGNMENT 7.1

ST7A. Use the above recipe to tessellate most of your screen. Apply each of the **Arrow** tools (**Translate**, **Rotate**, and **Dilate**) to your tessellation, just to observe the effects.

ST7B. Continuing, apply **Edit | Merge** to put your points *E* and *F* on small circles. Animate *E* and *F*, separately and together.

ST7C. Continuing, replace the two small circles with two short segments. Animate *E* and *F*, separately and together.

PROJECT 1: CONICS IN POLAR COORDINATES

The first part of this chapter introduces graphing, in polar coordinates, of functions of the form $r = r(t)$, where t is a directed angle. Among the curves which can be conveniently graphed in this manner are certain conics. Each nondegenerate conic has an eccentricity, which here and in Chapter 3 is denoted by e.

Recall that the definition of e is given in terms of a line D and a point F not on D. The locus of a point P for which the ratio $|PF|/|PD|$ stays constant is a conic; the constant is the eccentricity, e, and the conic is a circle, ellipse, parabola, or hyperbola according as $e = 0$, $0 < e < 1$, $e = 1$, or $e > 1$.

The functions given by

$$r = \frac{ed}{1 + e\cos\theta}$$

$$r = \frac{ed}{1 - e\cos\theta}$$

$$r = \frac{ed}{1 + e\sin\theta}$$

$$r = \frac{ed}{1 - e\sin\theta}$$

are conics, where $e > 0$ is the eccentricity and $|d|$ is the distance between the focus F at $(0, 0)$ and the corresponding directrix, D.

Part 1. Use Sketchpad's polar graphing capabilities to graph each of the following curves, along with a plot of point F and line D, and a printed measure of $|PF|/|PD|$ for several choices of P on the curve:

$$r = \frac{2}{1 + \cos\theta} \qquad r = \frac{1}{1 + 2\sin\theta}$$

$$r = \frac{2}{1 - \cos\theta} \qquad r = \frac{2}{2 - \sin\theta}$$

Part 2. Sketch a polar grapher for the equation

$$r = \frac{ed}{1 + e\cos\theta}$$

where e and d can be continuously varied by the user. Include a sketch of a focus and directrix, as well as a confirmation of the value of e by the ratio $|PF|/|PD|$.

PROJECT 2: AXES, VERTICES, AND FOCI OF CONIC THROUGH 5 POINTS

The objective of this project is to emulate Figure 8.19. Section 4 discusses the construction of the center of the conic determined by 5 points. Here, we wish to see how to construct other main features of that conic. The underlying geometry is presented in Dan Pedoe's "Pascal Redivivus: II", *Crux Mathematicorum* 5 (1979) 281–287. It is quite possible that you will need to have a copy of Pedoe's article in order to carry out this project.

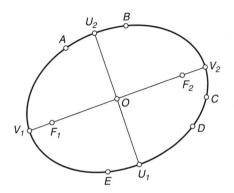

FIGURE 8.19 Ellipse of 5 given points A, B, C, D, E, with center O, vertices V_1, V_2, and foci F_1, F_2. If you drag A, B, C, D, E so that the curve is no longer an ellipse, it will still be a conic with center O. For some choices of A, B, C, D, E, technological limitations force the axes and V_1, V_2, F_1, F_2 to appear improperly.

Part 1. Emulate Figure 8.19.

Part 2. Using your sketch, confirm with measurements that your curve satisfies the defining equation in the caption of Figure 3.5.

Part 3. Select the conic of 5 given points and apply **Construct | Point on Object**. Label the point as P. Sketch the line T tangent to the conic at P. (Perhaps you have a tool named **conic tangent** from **ST4D**.) Confirm that the perpendicular to T at P bisects $\angle F_1PF_2$.

PROJECT 3: LINES TANGENT TO CONICS

Conics have found applications in many fields, especially optics and eye physiology. Among many ophthalmologists interested in conics is Dr. Lawrence Evans, who devised the sketches in **Parts 1** and **2** of this project.

Part 1. The objective here is to sketch the conic that passes through 3 independent points A, B, C and is tangent to lines BD and CE, where D and E are independent points. Here are the steps, adapted from Dr. Evans's sketch so that a tool can be readily created:

1. Sketch $F = BD \cap CE$.

2. Sketch $G = $ the rotation of point A through $93.728°$ about point B.

3. Sketch $H = AB \cap FG$.

4. Sketch $I = AC \cap FG$.

5. Sketch $J = $ the rotation of point B through $98.043°$ about point C.

6. Sketch $K = AB \cap FJ$.

7. Sketch $L = AC \cap FJ$.

8. Sketch $M = BI \cap CH$.

9. Sketch $N = BL \cap CK$.

(Actually, *G* and *J* could be replaced by many other points, but stipulating them specifically enables the creation of a simpler tool than would otherwise be the case.) Your sketch now has 5 points that lie on the desired conic. To sketch the conic, apply **conic thr 5 pts** to the selected points *A*, *B*, *C*, *M*, *N*. After checking that the resulting conic is the right one, create a tool **conic3P2T**. To do this, your selected objects should be *A*, *B*, *C*, *D*, *E*, *M*, *N* and the conic. The first five of these are givens and the remaining three are results. The reason for including *M* and *N* is that after applying **conic3P2T**, there will be 5, rather than 3, points on your conic available for selecting. This will be necessary, for example, if you wish to apply **conic center**.

Part 2. Here, we wish to sketch the conic that passes through 4 independent points *A*, *B*, *C*, *D* and is tangent to line *AE*, where *E* is an independent point.

1. Sketch $F = AB \cap CD$.

2. Sketch $G = $ the rotation of point *A* through $93.572°$ about point *B*.

3. Sketch $H = AE \cap FG$.

4. Sketch $I = AD \cap FG$.

5. Sketch $J = BI \cap CH$.

Your sketch now has 5 points that lie on the desired conic. To sketch the conic, apply **conic thr 5 pts** to the selected points *A*, *B*, *C*, *D*, *J*. After checking that the resulting conic is the right one, create a tool **conic4P1T**. To do this, your selected objects should be *A*, *B*, *C*, *D*, *J* and the conic.

Part 3. Use the tool **conic3P2T** from **Part 1** to sketch the conic determined as follows:

1. tangent to the *x*-axis at a movable point *B* on the *x*-axis

2. tangent to the *y*-axis at a movable point *C* on the *y*-axis

3. passes through a movable point *A* on the vertical line through point *B*

Apply **conic center** to your conic, and sketch the locus of the center as a function of *A*. Then sketch the locus of the center as a function of *C*.

Part 4. Use the tool **conic4P1T** from **Part 2** to sketch the conic determined as follows:

1. tangent to the *x*-axis at a movable point *A* on the *x*-axis

2. passes through independent points *B*, *C*, *D*

Apply **conic center** to your conic, and sketch the locus of the center as a function of *A*.

Part 5. Suppose *ABC* is a triangle. Figure out with pencil and paper how to construct the parabola tangent to side *AB* at *A* and tangent to side *CB* at *C*. Then create a sketch based on your construction.

PROJECT 4: A SPECIAL ANGLE

Consider the dashed segments in Figure 8.20. Each makes an angle with a nearby side of △*ABC*, and that angle has the same measure at all three vertices. As you survey from left to right, the angle in question increases. The resulting triangle inside △*ABC* becomes smaller and eventually shrinks to a point. When this happens, the special angle is known as the Brocard angle of the triangle.

Part 1. Start with an arbitrary triangle *ABC*. Elsewhere on your screen, sketch a circle. Label its center *O*, and put two movable points *D* and *E* on the circle. Drag *E* so that when you select *D*, *O*, *E* in that order, the angle *DOE* is about 5° and is oriented counterclockwise. Use this angle to rotate *B* about *A*, and *C* about *B*, and *A* about *C*. The result should resemble one of the configurations in Figure 8.20. Animate *E*.

FIGURE 8.20 Dashed lines making equal angles

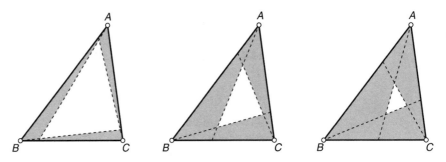

Part 2. Continuing, print a measurement of ∠*DOE*. Print measurements that confirm that the Brocard angle, customarily denoted by ω, satifies the following identity:

$$\cot \omega = \cot A + \cot B + \cot C$$

Also, label as *A′*, *B′*, *C′* the pairwise intersections of the rotated lines through *A*, *B*, *C*, respectively. Plot the loci of *A′*, *B′*, *C′* as functions of *E*.

Part 3. The continuous-action argument leading to **Part 1** is an interesting technique with applications in many places, but it isn't a Euclidean construction. The most common construction of the Brocard angle of △*ABC* is as follows:

Let O_A be the circle passing through *A* and tangent to line *BC* at *B*; similarly, let O_B be the circle passing through *B* and tangent to line *CA* at *C*; and let O_C be the circle passing through *C* and tangent to line *AB* at *A*. The three circles concur in a point, often called the *1st Brocard point* of △*ABC*. Denote it by *U*. The Brocard angle is ∠*UAB*.

Sketch *U* as just described. Confirm that ∠*UAB* = ∠*UBC* = ∠*UCA*, this being the Brocard angle.

Part 4. Emulate **Parts 1** and **2**, but this time, rotate clockwise: rotate *C* about *A*, *A* about *B*, and *B* about *C*. In the resulting configuration, as ∠*DOE* approaches the Brocard angle, the triangle *A′B′C′* shrinks to the *2nd Brocard point*.

PROJECT 5: MIQUEL LINES

In Section 6, we saw that for any two triangles *ABC* and *DEF*, infinitely many copies of *DEF* are inscribed in *ABC*. The copies are all similar to one another and are conveniently indexed and driven by a point *D* on line *BC*. The copies "twirl" about a certain point *P*; that is, they "rotate and change size but not shape".

Before sketching, use **Edit | Advanced Preferences | Sampling** to type 70 for the number of samples for new loci.

Part 1. Starting with **ST6A**, sketch the circumcenter *O* and the orthocenter *H* of △*DEF*. Then sketch the loci of *O* and *H* as functions of *D*. Determine whether these loci are unbounded lines or bounded segments.

Part 2. Continuing, sketch the locus of the Euler line of △*DEF* as a function of *D*. As always, print your observations.

Part 3. Sketch the circumcircle of △*DEF*. Then sketch its locus as a function of *D*.

Part 4. Sketch the incircle of △*DEF*. Then sketch its locus as a function of *D*.

PROJECT 6: THREE-DIMENSIONAL SKETCHING

Try this thought-experiment: you are standing still except for the index finger of your right hand, which is tracing a circle in the air. The circle is about one foot across, and your finger is going around it repeatedly. The plane in which you are tracing this circle is perpendicular to the direction you are facing. The motion of your finger represents "pure rotation".

Now, while continuing to trace the circle, slide forward at a uniform speed. The sliding represents "pure translation", and the locus of your nose is (ideally speaking) a line. But what about the locus of your finger? Its locus is a helix, perhaps the simplest *space-curve*—that is, a curve in three-dimensional space that cannot be confined to a plane. Even though this curve is a space curve, it can be represented on a two-dimensional Sketchpad screen, as in Figure 8.21.

Part 1. Figure out how to sketch, on Sketchpad, a helix such as in Figure 8.21. Then sketch it.

Part 2. A sphere can be effectively shown on a two-dimensional screen by a certain positioning of a circle and two ellipses. Do this on Sketchpad.

Before leaving the subject of space-curves, you might ask yourself, "What is the most common space curve"? The helix is often seen in the cord to a telephone receiver, but another common space curve, perhaps even more common than the helix—if you don't count DNA and electromagnetic radiation—is the one on baseballs, tennis balls, and basketballs. It seems strange that such a curve (which can be defined mathematically in terms of minimal-average-distance to points on a sphere) has no common name.

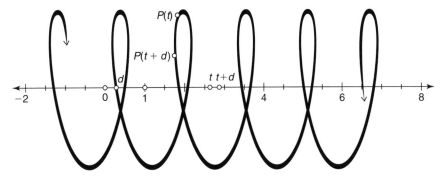

$$x_t = 2.64 \quad x_m \cdot x_t = 8.24$$
$$x_a \cdot (x_m \cdot x_t) + x_b \cdot \cos((x_m \cdot x_t)) = 1.86$$
$$x_c \cdot \sin((x_m \cdot x_t)) = 1.85$$

$x_a = 0.25$

$x_b = 0.54$

$x_c = 2.00$

$x_m = 3.12$

$$x_{t+d} = 2.91 \quad x_m \cdot x_{t+d} = 9.09$$
$$x_a \cdot (x_m \cdot x_{t+d}) + x_b \cdot \cos((x_m \cdot x_{t+d})) = 1.76$$
$$x_c \cdot \sin((x_m \cdot x_{t+d})) = 0.67$$

FIGURE 8.21 A helix

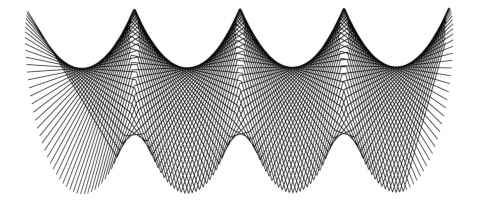

FIGURE 8.22 Pattern formed by reducing the number of samples and increasing d in the sketch that produced Figure 8.21

PROJECT 7: CIRCLE OF CURVATURE

For most of the curves Γ that you've heard of, there is, associated with each point P on Γ, a number called *curvature*. Curvature, often denoted by the Greek letter kappa, κ, measures the extent of curving at P. In other words, κ is a function of P, and we write it as $\kappa(P)$. If $\kappa(P)$ stays constant as P traverses the curve, then the curve is said to have constant curvature and is either a circle or part of a circle. If the constant is zero, then the circle has infinite radius and is a line.

The number $1/\kappa$ is called the *radius of curvature*. It, too, of course, is a function of P. If Γ is a circle, then $1/\kappa$ is a constant, and in fact it equals the radius of the circle. Now suppose Γ is the graph of a polynomial or a trig function, and P is a point on Γ. Let T be the line tangent to Γ at P. The line through P perpendicular to Γ is called the *normal line* at P. If P is not an inflection point of Γ, then at P the curve is concave in one of the directions proceding from P along the normal line. Choose that direction and let O be the point at distance $1/\kappa$ from P. The point O is the *center of curvature*. The circle $\circ(O, |OP|)$ is the *circle of curvature*, also called the *osculating circle*. The verb *to osculate* means *to kiss,* in agreement with the fact that this circle is tangent to Γ at P. In other words, this circle, in a limiting sense, more closely approximates the curve at P than any other circle.

The radius of curvature, r, at a point (x, y) on the parabola $y = ax^2 + bx + c$ is given by

$$r = \frac{[1 + (2ax + b)^2]^{3/2}}{|2a|}$$

This formula appears in Sketchpad notation in Figure 8.23. Imagine, when viewing Figure 8.23, that point P is moving on the parabola. The accompanying circle may be imagined to roll on the inside of the parabola, with radius gradually increasing as P moves away from the vertex of the parabola. When P is the vertex, the circle has its least possible radius.

Emulate Figure 8.23, starting with a parabola grapher as in Figure 3.13. Create loci as in Figure 8.24 by trying various number of samples chosen from **Edit | Advanced Preferences.**

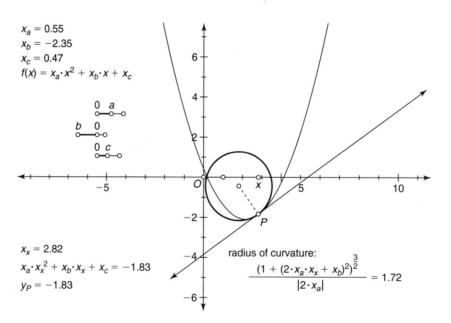

$x_a = 0.55$
$x_b = -2.35$
$x_c = 0.47$
$f(x) = x_a \cdot x^2 + x_b \cdot x + x_c$

$x_x = 2.82$
$x_a \cdot x_x^2 + x_b \cdot x_x + x_c = -1.83$
$y_P = -1.83$

radius of curvature:
$$\frac{(1 + (2 \cdot x_a \cdot x_x + x_b)^2)^{\frac{3}{2}}}{|2 \cdot x_a|} = 1.72$$

FIGURE 8.23 Circle of curvature tangent to the parabola $y = ax^2 + bx + c$ at P. Drag x to move P. Use the sliders to vary a, b, and c.

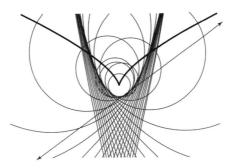

FIGURE 8.24 Loci of circle of curvature and its center for the curve in Figure 8.23

As indicated in Figure 8.24, the locus of the center of the circle of curvature is a curve. You've probably gleaned from your curve-sketching activities that there are hundreds of named curves and millions of unnamed curves. For more information about specific curves—or just to have a good time, visit Xah Lee's website, A Visual Dictionary of Special Plane Curves: **http://www.xahlee.org/SpecialPlaneCurves_dir/ specialPlaneCurves.html**

CHAPTER **NINE**

Hyperbolic Geometry

VIRTUALLY EVERYONE LIVING two centuries ago believed there was only one kind of geometry. This geometry, they believed, matches physical reality. They imagined no other possibility. However, a certain feature of Euclidean geometry became a seed for the discovery of other geometries that match certain characteristics of the physical universe better than Euclidean geometry does. That seed is Euclid's "fifth postulate", commonly called the "parallel postulate". It is equivalent to the following statement:

> If L is a line and P a point, then there exists only one line through P parallel to L.

After centuries of trying, no one was able to derive this simple-looking proposition from the other Euclidean postulates. Consequently, the possibility that this postulate is *independent* of the others—meaning that the others don't imply it—motivated a new kind of search. Mathematicians realized that new geometries may exist in which the "physically obvious" parallel postulate is replaced by some other, and that mathematical truth, instead of being absolute, depends on what postulates (or axioms) are assumed.

At this point, we should clarify what is meant by "a geometry". A geometry is *a collection of postulates about points, lines, circles (and perhaps other fundamental notions) together with all propositions that are implied by those postulates.* The collection must be consistent; that is, if a proposition is implied, then its negation must not be implied. The distinguishing feature of a postulate is that it is so basic that it cannot be derived from "more basic" propositions—because there are no "*more* basic" propositions. (If there were, *they* would be the postulates.)

The search for new geometries, then, was driven by the possibility that Euclidean geometry, and in particular, the parallel postulate, is not the "whole truth". One of the earliest replacement-postulates to be considered, and the first to pay off, is this:

> If *L* is a line and *P* a point, then there exists more than one line through *P* parallel to *L*.

A very interesting geometry springs from this postulate. In this chapter, we shall use Sketchpad to sample it. It is called "hyperbolic geometry", a name given in 1871 by Felix Klein to geometries introduced by J. Bolyai and N. I. Lobachevsky in the early 1800s. (Klein named other geometries "elliptical" and "parabolic".)

Historians of mathematics, and also historians of physics, astronomy, and philosophy, have crafted wonderful paragraphs to portray the revolutionary effect of the discovery of noneuclidean geometries. Such paragraphs are often quoted. For example, Howard Eves writes in *An Introduction to the History of Mathematics*:

> The creation of the Lobachevskian geometry not only liberated geometry but had a similar effect on mathematics as a whole. Mathematics emerged as an arbitrary creation of the human mind and not as something essentially dictated to us of necessity by the world in which we live. The creation of the non-Euclidean geometry, by puncturing a traditional belief and breaking a centuries-long habit of thought, dealt a severe blow to the absolute truth viewpoint of mathematics. In the words of Georg Cantor, "the essence of mathematics lies in its freedom".

Hyperbolic geometry is represented by various models. One of them is especially well suited to Sketchpad-based exploration. Introduced by Henri Poincaré, it is called the Poincaré disk model.

SECTION 1 # The Poincaré Disk Model

It is not part of the definition of "line" that it must be "straight". In fact, in geometric postulates, the term "line" is an undefined term, and "straight" isn't even mentioned. The first thing about hyperbolic geometry that catches the eye is that lines are *not* "straight". Usually, when someone says "parallel lines", the mind's eye pictures two straight lines separated by a fixed distance. However, the definition of parallel doesn't really depend on straightness. Instead, lines are defined to be parallel if they do not meet. Since the underlying distinguishing feature of a hyperbolic geometry involves more than one parallel line where common sense says there is only one, let's begin our Sketchpad-based exploration with a sketch that could be named **many parallels**.

To do so, we need a Sketchpad document that enables Sketchpad-like constructions of hyperbolic lines. For this purpose, we'll use the file named **Poincare Disk.gsp**, which you can find on your copy of the CD entitled *Geometry in Action: Selected Sketches*.

To begin, apply Sketchpad's **File | Open | Poincare Disk**. The resulting screen will include a circle, which we'll call the *Poincaré circle*. It is the interior of this circle (without the boundary) that is meant by "Disk" in the name Poincaré Disk. Click on **Disk Controls**, and you'll see points appear, one at the center of the disk and the other on the circle.

Show the label for the center of the disk; it is **P.Disk Center**. Henceforth in this chapter this special point will be referred to as **O**. It would be wise to change the label to **O** on the screen, in order to match your labels with this discussion. Next, show the label for the other point; it is **P.Disk Radius**. Relabel this point as **R**. Drag it to see that it determines the radius of the disk.

In order to perform hyperbolic constructions, click the **Custom Tools** icon in Sketchpad's main toolbox to see a list of tools:

Hyperbolic Segment

Hyperbolic Line

Hyperbolic P. Bisector

Hyperbolic Perpendicular

Hyperbolic A. Bisector

Hyperbolic Circle by CP

Hyperbolic Circle by CR

Hyperbolic Angle

Hyperbolic Distance

Use the **Point Tool** in Sketchpad's main toolbox to sketch two points, labeled A and B. Click the **Custom Tools** icon again. In the resulting dialog box, click **Hyperbolic Line**. Then, in the dialog box, click **Show Script View** and see another dialog box. If any objects on your screen are selected, de-select them.

Let's take a good look at this **Show Script View** box. At the top, you see the name of the tool, **Hyperbolic Line**, followed by the word **Script**, meaning a written version of the construction that will be performed by the tool. To see the whole script, you can use the two scroll bars in the dialog box, or, better, you can make a printout on paper by right-clicking on the interior of the dialog box and then clicking on **Print**.

The interior of the dialog box consists of sections headed

Assuming: Given: and **Steps:**

Now, and in the future every time you apply any of the hyperbolic tools listed above, you must select, in order, O, then R, and then other points as required for the tool.

So, in order to sketch the hyperbolic line of A and B, select O, R, A, B. As you do this, notice that the **Show Script View** box shows your selections. When you've finished selecting, the box will indicate two opportunities after **Apply**. Click the second one, labeled **All Steps**. You'll see a quick construction, after which all its objects, except the final one, will be hidden. The final object is the hyperbolic line of the points A and B. For any points X and Y we'll use the notation hXY for the hyperbolic line passing through X and Y.

When using **Poincare Disk**, feel free to hide and move captions. For example, in Figure 9.1, the explanatory paragraph has been removed, and the button labeled **Disk Controls** has been moved (by placing the cursor on the left edge of the button before dragging).

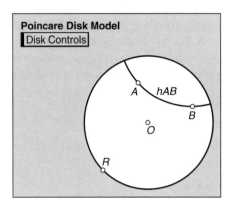

FIGURE 9.1 The points O, R, and a hyperbolic line hAB

HG1A. Follow the above discussion with a sketch using **Poincare Disk**. Then add two more points to the disk, labeled C and D, and sketch hCD. If it meets hAB, then drag C and D to positions such that the two lines do not meet; i.e., so that hCD is parallel to hAB. Then sketch points E and F such that hEF misses both hAB and hCD. Explain in a caption what is illustrated here that distinguishes hyperbolic geometry from Euclidean geometry. (It is this sketch that could be named **many parallels**.)

HG1B. Let's define straightness for a line as follows: a hyperbolic line hXY is *straight* if it lies in the Euclidean line XY. Discover experimentally all conditions under which hXY is straight. Write your discovery in a caption.

HG1C. One very attractive feature of **Poincare Disk** is that Euclidean constructions can be performed, in addition to using hyperbolic tools. Returning to **HG1A**, sketch a Euclidean line through O. Label it L. Use **Transform | Reflect** to reflect hAB in L. Call the resulting object H. Predict the reflection of H in L, and then confirm or refute your prediction.

HG1D. Starting with **HG1B**, delete all objects except O and R. Select O and apply **Graph | Define Origin**. You'll see the usual x- and y-axes. Label as N, W, S, E the points in which these axes meet the Poincaré circle. The 4 points, taken 2 at a time, determine a total of 6 hyperbolic lines. Likewise, they determine 6 Euclidean lines. Sketch all 12 lines. But before you do, convince yourself that $hNS \neq NS$. (You may not be able to use points right on the circle; in that case, use points inside the circle that are visually very close to N, W, S, E.)

Arcs and Angles

Compared to other models of hyperbolic geometry and other noneuclidean geometries, two very useful features of the Poincaré disk are that it is a subset of the Euclidean plane and that hyperbolic lines are Euclidean circular arcs. The first of these makes it possible to apply Sketchpad's Euclidean constructions to hyperbolic objects, as in **HG1C**. The second feature suggests a fundamental question: of all the circles through two given points A and B, from which one do we "borrow" to have the hyperbolic line hAB? The answer is quite simple: the circle that meets the Poincaré circle in 90°. Take a glance at any of the lines in **HG1A–D** to confirm, at least visually, that each hyperbolic line does indeed meet the Poincaré circle in 90°.

However, there is a question of definition here: 90° is an angle measurement, and the word "angle" ordinarily applies to a point where two Euclidean (i.e., "straight") lines meet. What is meant by the "angle" between two hyperbolic lines—that is, the "angle" between two circular arcs? The answer is this: the angle between the Euclidean lines tangent to the two arcs. Project 1 will explore this matter further. For now, it nice to know that the tool **Hyperbolic Angle** measures the angle between two hyperbolic lines. This is done by selecting, in order, three points A, B, C positioned so that the two hyperbolic lines are hAB and hAC.

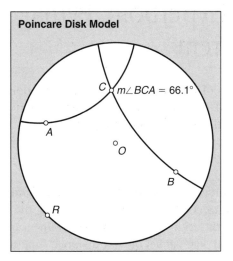

Poincare Disk Model

$m\angle BCA = 66.1°$

FIGURE 9.2 Angles between lines

ASSIGNMENT 2.1

HG2A. Sketch a hyperbolic triangle with vertices labeled A, B, C. Use **Hyperbolic Angle** to print measures of the hyperbolic angles CAB, ABC, and BCA. Then apply **Measure | Calculate** to print the sum, S, of these three angles. Drag points A, B, C around in an effort to determine greatest possible and least possible values of S. In a caption, print a sentence that begins like this: The sum, S, of the angles of a triangle appears to be less than 180°; to sketch a triangle with $S > 178°$, [finish the sentence]. In another caption, print a sentence regarding small values of S.

HG2B. Continuing, delete captions, and drag A, B, C so that angle ABC has a measure of 90°. Then drag B slowly to a position outside the Poincaré disk. Notice that lines hAB and hBC disappear as soon as B leaves the disk; this is appropriate, since B is outside the "space" of this geometry—in much the same way that a point above a horizontal Euclidean plane is outside the plane. Now drag B back into the Poincaré disk, but just barely. Try to drag A and C to positions for which angle ABC has a measure of 90°. Experiment, and then print a caption that explains what you decide can be said about any angle that has a vertex very close to the boundary of the Poincaré disk.

HG2C. In a Poincaré disk, sketch a movable Euclidean equilateral triangle ABC and printed measures of the Euclidean vertex angles. Then sketch the hyperbolic triangle ABC and use **Hyperbolic Angle** to print measures of the three hyperbolic vertex angles. Are they equal? Sketch a second hyperbolic triangle whose vertices are independent points D, E, F. Can you drag these points to positions in which triangle DEF has equal hyperbolic angles?

SECTION 3 # The Hyperbolic Pythagorean Theorem

The length of a segment is given by a function whose domain is the set of all pairs of points and whose image is often called "distance". The usual Euclidean distance function for points $P = (x_1, y_1)$ and $Q = (x_2, y_2)$ is given by

$$|PQ| = \sqrt{(x_1 - x_2)^2 + (y_1 - y_2)^2}$$

Of course, you've applied this familiar distance function in many ways in mathematics courses. In the Poincaré disk, the distance between P and Q is given by a different function. We'll examine it in Section 6. Here in Section 3, we'll simply use values of that distance function as calculated by **Hyperbolic Distance**.

The hyperbolic functions are presented in Project 5 of Chapter 5. Recall that they belong to the hyperbola $x^2 - y^2 = 1$ in the same way that the trig functions belong to the circle $x^2 + y^2 = 1$. In this section, we shall need one of the hyperbolic functions, namely the hyperbolic cosine, given by

$$\cosh x = (e^x + e^{-x})/2$$

The functions e^x and e^{-x} are easily accessible on Sketchpad's calculator.

Now suppose that *ABC* is a hyperbolic triangle in the Poincaré disk. Let *a*, *b*, *c* denote the hyperbolic lengths of the sides opposite angles *A*, *B*, *C*, respectively, as in Figure 9.3.

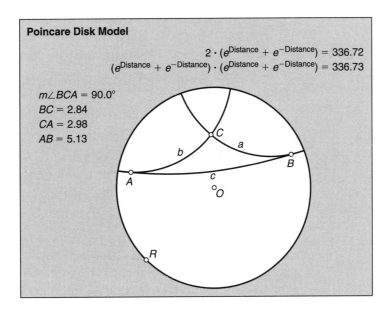

FIGURE 9.3 A hyperbolic right triangle

The Hyperbolic Pythagorean Theorem follows: *If ABC is a hyperbolic right triangle with right angle at C, then* $\cosh c = \cosh a \cosh b$.

ASSIGNMENT 3.1

HG3A. Sketch a hyperbolic right triangle *ABC*, and use the tool **Hyperbolic Distance** to confirm the Hyperbolic Pythagorean Theorem with printed measurements. (For reasons to be discussed in Section 6, it is sufficient to obtain measurements that are *nearly* equal.)

HG3B. Continuing, figure out how to animate a point *C′* in such a way that its children include points *A′* and *B′* such that *A′*, *B′*, *C′* are vertices of a hyperbolic right triangle. Then make it happen.

SECTION 4 The Hyperbolic Laws of Sines and Cosines

We begin by recalling the Law of Sines: *in a Euclidean triangle ABC with matching sidelengths a, b, c:*

$$\frac{\sin A}{a} = \frac{\sin B}{b} = \frac{\sin C}{c}$$

Analogous to this theorem is the following theorem:

Hyperbolic Law of Sines: *If UVW is a hyperbolic triangle and u, v, w are the hyperbolic sidelengths opposite hVW, hWU, hUV, respectively, then*

$$\frac{\sin U}{\sinh u} = \frac{\sin V}{\sinh v} = \frac{\sin W}{\sinh w}$$

For purposes of Sketchpad measurements, the following identity is useful: $\sinh x = (e^x - e^{-x})/2$.

There are two hyperbolic analogues for the Law of Cosines. Continuing with $\triangle UVW$, they are as follows:

Hyperbolic Law of Cosines for Sides:
$\cosh w = \cosh u \cosh v - \sinh u \sinh v \cos W$

Hyperbolic Law of Cosines for Angles:
$\cos W = -\cos U \cos V + \sin U \sin V \cosh w$

By putting $W = 90°$ in the first of these two, note that what remains is the Hyperbolic Pythagorean Theorem. Thus, this Hyperbolic Law of Cosines is stronger than the Hyperbolic Pythagorean Theorem, in the same way that the Law of Cosines is stronger than the Pythagorean Theorem.

ASSIGNMENT 4.1

HG4A. Sketch a hyperbolic triangle UVW, and use **Hyperbolic Distance** to confirm the Hyperbolic Law of Sines with printed measurements.

HG4B. Sketch a hyperbolic triangle UVW, and confirm the Hyperbolic Law of Cosines for Sides with printed measurements. Do this in one of the three ways indicated here:

$$\cosh u = \cosh v \cosh w - \sinh v \sinh w \cos U$$

$$\cosh v = \cosh w \cosh u - \sinh w \sinh u \cos V$$

$$\cosh w = \cosh u \cosh v - \sinh u \sinh v \cos W$$

After all the measurements are printed, determine, by dragging the vertices U, V, W, conditions under which certain measurements are not very accurate (because of computer limitations).

HG4C. Sketch a hyperbolic triangle UVW, and confirm the Hyperbolic Law of Cosines for Angles (cf. **HG4B**).

SECTION 5 # Inversion in a Circle

Recall from Chapter 3 that the inverse of a point P in a circle $\circ(O, r)$ is the point Q on line OP satisfying $|OQ| \cdot |OP| = r^2$, assuming that $P \neq O$. In Figure 9.4, imagine that the circle is the Poincaré circle, and anticipate that

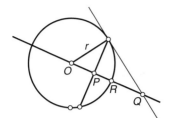

FIGURE 9.4 Inverse Q of a point P in $\circ(O, r)$

the model of hyperbolic geometry called the Poincaré disk depends largely and simply on inverses of Euclidean lines in the circle $\circ(O, r)$.

Figure 9.4 and the equation $|OQ| \cdot |OP| = r^2$ indicate that Q can be sketched by applying **Transform | Dilate**. Specifically, dilate R from center O with ratio $|OR|/|OP|$ (as in **LO3C**). Let's call the resulting tool **invert point**.

Our main objective in this section is to construct the hyperbolic line of two points A and B. There is a certain circle Λ that overlaps $\circ(O, r)$, and the required hyperbolic line hAB is that part of Λ that lies inside $\circ(O, r)$. The circle Λ is the one that passes through the points A, B, and their inverses in $\circ(O, r)$. We need only three of these four points to construct Λ. Here's the plan: apply **invert point** to construct the inverse, D, of A, as in Figure 9.5. The perpendicular bisectors of segments AD and AB meet in the center, U, of Λ, so that $\Lambda = \circ(U, |UA|)$.

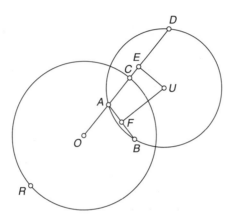

FIGURE 9.5 Circle Λ through points A, B, D

Next, label as A' and B' the points of intersection of the two circles, as in Figure 9.6. Apply **Construct | Arc Through 3 Points** to see the Euclidean circular arc that serves as a definition of hyperbolic line.

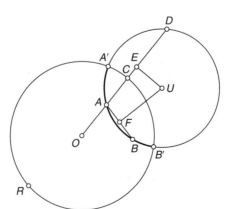

FIGURE 9.6 Hyperbolic line AB, alias the Euclidean arc $A'ABB'$

ASSIGNMENT 5.1

HG5A. Using the Poincaré circle as $\circ(O, r)$, emulate Figure 9.6. In a caption, explain the behavior of U as a function of A. Print a measure of the Euclidean arclength of segment hAB, as well as the hyperbolic length of segment hAB as given by **Hyperbolic Distance**. The two measures are not generally equal. This suggests that the hyperbolic distance function differs from the familiar Euclidean distance function. The former will be the subject of Section 6. (For this assignment, you may wish to select point O twice but find that you can't. In that case, use a nearby point O' instead, and later, drag it onto O.)

HG5B. Continuing, confirm by additional sketching that the inverse of point B lies on circle Λ.

HG5C. The angle between two circles at a point P where they meet is defined as the angle between the tangent lines to the circles at P. Confirm that the angle at point A' in **HG5A** is a right angle.

SECTION 6 # The Hyperbolic Distance Function

As indicated in **HG5A**, the hyperbolic distance function is quite different from the ordinary Euclidean distance function. With reference to Figure 9.6, the hyperbolic distance between the points AB is given by

$$H(A, B) = \left| \ln \left(\frac{|AB'|}{|BB'|} \cdot \frac{|BA'|}{|AA'|} \right) \right|$$

What right, you might ask, did Poincaré have to use a different "distance function"? The answer stems from an examination of what is meant by "distance". This will be considered further in Project 2.

For now, the appropriate point of view is that this new distance function fits a geometry determined by a collection of postulates. For the case at hand, the particular geometry is hyperbolic geometry. Other geometries are modeled by other distance functions—hundreds of them.

In previous sections, it has been noted that Sketchpad doesn't measure certain hyperbolic distances accurately. The reason for this can now be gleaned from the formula for $H(A, B)$. As A approaches the boundary of the Poincaré disk, the denominator $|AA'|$ approaches 0, and similarly for B. Consequently, if either point is close to that boundary, the resulting measurement of $H(A, B)$ may be inaccurate.

In fact, as $|AA'|$ approaches 0, the quotient $|BA'|/|AA'|$ grows without bound, and consequently, so does $H(A, B)$. That is to say, the distance function H is unbounded (whereas the usual Euclidean distance function, d, is bounded on a disk of radius r, since $d(P, Q) < 2r$ for every pair of points P and Q inside the disk). Sometimes, people call H and hyperbolic lines "infinite" when they should instead say "unbounded". (True, lines *are* infinite, but so are tiny segments; for example, the interval from 0 to one-billionth contains infinitely many points $1/n$, for $n \geq 10^9$ and many others.)

HG6A. Continuing from **HG5A**, use **Measure | Calculate** to print measurements needed to check that the formula for $H(A, B)$ agrees with Hyperbolic Distance.

HG6B. Create a sketch that confirms that for any points A and B in the Poincaré disk, there exists (1) a Euclidean reflection that carries A and B to points C and D such that $H(C, D) \neq H(A, B)$; and (2) a reflection that carries A and B to points E and F such that $H(E, F) = H(A, B)$.

HG6C. Create a sketch that confirms that for any points A and B in the Poincaré disk, there exists (1) a rotation that carries A and B to points C and D such that $H(C, D) \neq H(A, B)$; and (2) a nontrivial rotation that carries A and B to points E and F such that $H(E, F) = H(A, B)$.

SECTION 7 Hyperbolic Circles

Suppose $r > 0$ and U is a point in the Poincaré disk. The locus of a point P whose hyperbolic distance from U is r is called the *hyperbolic circle* with center U and radius r. Thus, hyperbolic circles are analogous to Euclidean circles. In keeping with this analogy, there is a unique hyperbolic circle having a given center and pass-through point.

Section 1 includes a list of tools that come with **Poincare Disk**. Two of them are **Hyperbolic Circle by CP** and **Hyperbolic Circle by CR**. The first sketches the circle having center C and passing through point P. The second sketches the circle having center C and radius R given by selecting two points separated by hyperbolic distance R.

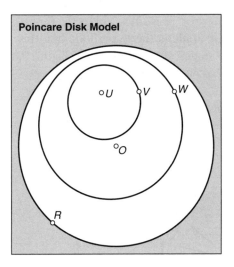

FIGURE 9.7 Concentric hyperbolic circles

HG7A. In the Poincaré disk, animate a point P on a circle sketched by **Hyperbolic Circle by CP**, with a printed measure of the hyperbolic distance $H(P, C)$, where C is the center of your circle. Then sketch, as a locus, the hyperbolic midpoint, M, of hyperbolic segment hPC, with printed measures of the hyperbolic lengths of segments hMC and hMP. Finally, sketch the hyperbolic circle that has diameter hPC.

HG7B. In the Poincaré disk, sketch an arbitrary hyperbolic circle, and let r denote its hyperbolic radius. Then its hyperbolic circumference and area are $2\pi \sinh r$ and $4\pi \sinh^2(r/2)$, respectively. Print these measures, and print what is strikingly different about these, in contrast to the Euclidean circumference and area for the same circle.

HG7C. In the Poincaré disk, let U and V be independent points, and let $r = H(U, V)$. Sketch the hyperbolic $\circ(U, r)$. Let W be an independent point outside $\circ(U, r)$ and let $r' = H(U, W)$. Sketch the hyperbolic $\circ(U, r')$. Let P be a movable point on $\circ(U, r')$. Sketch the hyperbolic line hUP and a point P' in which this line meets $\circ(U, r)$. Print measurements of

$$H(U, P) \quad H(U, P') \quad H(P, P')$$

In a caption, print a simple equation (or two equations) for $H(P, P')$ in terms of $H(U, P)$ and $H(U, P')$. Animate P and check whether the printed measurements support your equation(s).

SECTION 8 Hyperbolic Triangle Centers

One of the four ancient Greek centers of a Euclidean triangle is the incenter, I. Recall that I is the point where the bisectors of vertex angles A, B, C meet. As the Hyperbolic Tools listed in Section 1 include one named **Hyperbolic A. Bisector**, it is natural to ask whether there is a hyperbolic incenter. This question presumes, correctly, that the hyperbolic angle bisector of a hyperbolic angle is the hyperbolic line that separates the angle into two hyperbolic angles of equal measure. Figure 9.8 depicts three such bisectors and a hyperbolic incenter.

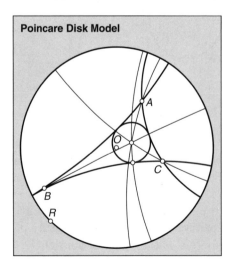

FIGURE 9.8 Hyperbolic incircle

The name "incenter" for I corresponds to the fact that this point is the center of the circle inscribed in $\triangle ABC$. Does this carry over to hyperbolic geometry? Figure 9.8 suggests that the answer is yes.

ASSIGNMENT 8.1

HG8A. Emulate Figure 9.8. Then drag points so that the Poincaré disk is relatively much larger than $\triangle ABC$. After experimenting, describe, in a caption, evidence that as the Euclidean radius of the Poincaré disk increases, any fixed hyperbolic configuration within that disk comes closer to matching a certain Euclidean configuration.

HG8B. Construct and save a tool that deserves the name **hyperbolic circle thr 3 pts**.

HG8C. Create a sketch that confirms or challenges the notion of "hyperbolic centroid" of a hyperbolic triangle.

HG8D. Create a sketch that confirms or challenges the notion of "hyperbolic orthocenter" of a hyperbolic triangle.

PROJECTS

PROJECT 1: ANGLE BETWEEN HYPERBOLIC LINES

Section 2 discusses the angle between two hyperbolic lines, defined as the angle between two Euclidean lines.

Part 1. In the Poincaré disk, sketch hyperbolic lines hAB and hCD using independent points A, B, C, D. Drag these points so that the lines meet in a point; label it P. Figure out how to draw Euclidean lines through P, one of them tangent to hAB and the other tangent to hCD. Print a measure of the acute angle between your Euclidean lines. Use **Hyperbolic Angle** to print a measure of the hyperbolic angle between hAB and hCD. Something is amiss if the two measures don't stay (nearly) equal as you drag the points A, B, C, D.

Part 2. Recall that hAB is an arc of a Euclidean circle. Complete this circle, and complete the circle of which hCD is an arc. Label as P' the point where the circles meet outside the Poincaré disk. Sketch the tangent lines at P' and print the measure of the acute angle between them. In accord with **L03F**, this angle equals the angle measured in **Part 1**; this is important because this present angle is the "original"—a parent of the angle between hAB and hCD.

Part 3. A hyperbolic triangle with hyperbolic sidelengths a, b, c is called *equilateral* if $a = b = c$. Let A be an independent point in the Poincaré disk. Sketch an equilateral hyperbolic triangle, one of whose vertices is A. Print measurements of the sidelengths and vertex angles of your hyperbolic triangle.

Part 4. A hyperbolic quadrilateral with hyperbolic sidelengths a, b, c, d is called a *square* if $a = b = c = d$. Let Λ be a circle and A be a movable point on Λ. Sketch a square, one of whose vertices is A. Print measurements of the sidelengths. Animate A.

PROJECT 2: WHAT IS A DISTANCE FUNCTION?

Let's abstract what makes the usual kind of distance "work", and then decree that any function that has the same abstract properties will be called a distance function. The three properties that "work" are typified by these observations:

1. (distance from Omaha to Omaha) $= 0$

2. (distance from Omaha to Denver) = (distance from Denver to Omaha)

3. (distance from Omaha to Tulsa) $\leq$ (distance from Omaha to Denver) + (distance from Denver to Tulsa)

As a first level of abstracting, observation (1) becomes

> dist from a point to itself is 0.

Observation (2) becomes

> dist from P to Q equals dist from Q to P.

Observation (3) becomes

> (dist from P to R) $\leq$ (dist from P to Q) + (dist from Q to R).

As a second level of abstracting, a function d, whose domain is a set of pairs of points, is a *distance function* (or *metric*) if for *all* choices of points $P, Q, R,$ the following hold:

(i) $d(P, P) = 0$

(ii) $d(P, Q) = d(Q, P)$

(iii) $d(P, R) \leq d(P, Q) + d(Q, R)$

When checking whether a given function d is a distance function, it is often easier to verify (i) and (ii) than (iii). The latter is called the *triangle inequality*. Consider Omaha, Denver, and Tulsa: (iii) matches the fact that the shortest distance from Omaha to Tulsa is on a line; if instead you go first to Denver, you've created a triangle, and your distance going along the other two sides is greater than if you just stay on the side straight from Omaha to Tulsa—hence the name triangle inequality.

It is a worthwhile exercise to verify, on your own, that (i)-(iii) hold for ordinary Euclidean distance and for the hyperbolic distance function presented in Section 6.

We turn next to an alternative distance function for the Euclidean plane. It is sometimes called the *taxi distance,* or *taxicab metric.* Imagine that you are in a western town where the streets form a north-south and east-west square grid. In order to get from one intersection to another, you must follow the grid—you cannot cut across. Here, the "distance" between points means the shortest path-length between the points, using a path that is confined to the streets.

Let's extend the taxi distance to the Euclidean plane by extending the square grid to the xy-coordinate system. Then east-west paths correspond to paths parallel to the x-axis, and north-south, to the y-axis. The taxi distance function is then expressed in terms of points $P = (x_1, y_1)$ and $Q = (x_2, y_2)$ by

$$D(P, Q) = |x_1 - x_2| + |y_1 - y_2|$$

Part 1. Let P, Q, R be independent points. Confirm with printed measurements that

$$D(P, R) \leq D(P, Q) + D(Q, R)$$

when P, Q, R are dragged.

Part 2. Use Sketchpad to sketch a "circle" using D as the distance function. That is, for your choice of a point P and radius r, sketch all the points Q satisfying $D(P, Q) = r$. Then sketch all the points Q satisfying $D(P, Q) \leq r$.

Part 3. Verify with two such "circles" of different sizes can meet in more than two points.

PROJECT 3: TWO POINCARÉ DISKS

It can be enlightening to compare configurations in two separate Poincaré disks. You can carry out a sketch on one disk, and then apply **Edit | Copy** and **Edit | Paste** to obtain a replica. Then you can work on both disks, independently!

The two-disk figures below were created in that way. The left disk shows a configuration ready for animation of P. The right disk shows a locus that is a function of P.

Part 1. Emulate Figure 9.9.

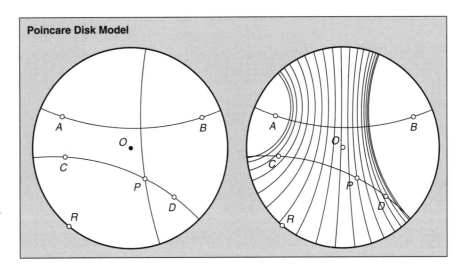

FIGURE 9.9 The perpendicular to hAB from P, ready for animation of P on hCD. On the right, the locus of the perpendicular as P moves on hAB.

Part 2. Emulate Figure 9.10.

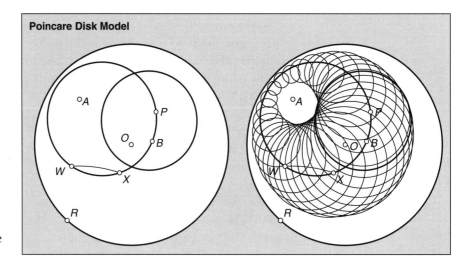

FIGURE 9.10 The circle with center P passing through Q, ready for animation of P. On the right, the locus of the circle as P moves on hAB.

Part 3. Create another two-disk sketch, the left part for animation and the right one to show a locus together with an animation of the locus. One possibility is the locus of a hyperbolic circle tangent to a fixed hyperbolic circle at a point T. If you create such a sketch, be sure to animate T to see some striking action.

Appendix

Assignments by Filename
Not Including End-of-Chapter Projects

FD7D	Distance between incenter and circumcenter	
FD7E	Locus of intersecting tangent lines	
FD7F	Orbits of excenters	
FD7G	Intersecting orbits	
FD8A	Loci of circumcenter and circumcircle	
FD8B	Similar intermediate triangles	
FD8C	Perspective initial triangles	
FD8D	A triangle and a point	
FD9A	Monge's theorem: external tangents	
FD9B	Internal tangents	
FD9C	Special cases for external tangents	
FD9D	Intersections of internal tangents	

Eight   Selected Topics

ST1A	Spiral driven by point on circle
ST1B	Spiral driven by point on line
ST2A	Graphs in polar coordinates: circles and a spiral
ST2B	Graphs: spirals
ST2C	Graphs: cosine circles
ST2D	Graphs: sine circles
ST2E	Graphs: cosine roses
ST2F	Graphs: sine roses
ST2G	Roses controlled by sliders
ST3A	Graphs: cosine limaçons
ST3B	Graphs: sine limaçons
ST3C	Limaçons controlled by sliders
ST3D	Geometric sums
ST3E	Cardioidal midpoint locus
ST4A	Conic center
ST4B	Poncelet orbit of conic center
ST4C	Conic with four points on a circle
ST4D	Line tangent to a conic
ST5A	Five-pointed star
ST5B	Golden mean as a length
ST5C	Odom's construction
ST5D	Regular pentagon and golden mean

Index